GEOL STRUCTURES

GEOLOGICAL STRUCTURES

An Introductory Field Guide

Chris and Helen Pellant

BLOOMSBURY
LONDON • OXFORD • NEW YORK • NEW DELHI • SYDNEY

Bloomsbury Natural History
An imprint of Bloomsbury Publishing Plc

50 Bedford Square
London
WC1B 3DP
UK

1385 Broadway
New York
NY 10018
USA

www.bloomsbury.com

BLOOMSBURY and the Diana logo are trademarks of Bloomsbury Publishing Plc

First published by Bloomsbury, 2016

British Library Cataloguing-in-Publication Data
A catalogue record for this book is available from the British Library.

Library of Congress Cataloguing-in-Publication data has been applied for.

ISBN: PB: 978-1-4729-2726-2
ePDF: 978-1-4729-4349-1
ePub: 978-1-4729-2727-9

2 4 6 8 10 9 7 5 3 1

Designed and typeset in UK by Susan McIntyre
Printed by RR Donnelly

CONTENTS

Jointed Pre-Cambrian sandstone, Sutherland, seen in a sea stack, probably formed by erosion along joints.

INTRODUCTION

The term 'geological structures' encompasses a wide range of geological features and can be difficult to define. Structural geology is the scientific study that deals, in the main, with the deformation and breaking of rocks, considering folding and faulting and their associated features. However, the term 'geological structures' covers a much wider field.

Structures are found within rocks and on rock surfaces as patterns, shapes and lines running on and through them. There is a great range in the size of geological structures, from a vast feature such as the San Andreas Fault to minute crenulations on slate. Each of the three main groups of rocks has its own suite of structures. Igneous rocks exhibit features such as sills and dykes, pillow lavas and columnar jointing; metamorphic rocks may be affected by cleavage and schistosity; sedimentary rocks contain a wide range of bedding structures, desiccation cracks, ripple marks and structures of chemical and biogenic origin.

Landforms are elements in the make-up of the Earth's surface, often related to rock structure and the way this has influenced the creation of surface features through erosion and weathering. Geomorphology is the science concerned with the understanding of landforms. It considers how the underlying rocks and their structure and composition have affected and produced the landscape.

This book is intended to be an introduction to geological structures and hopes to explain some of the more common types. A number of examples from different locations are illustrated, to show variations and further help understanding. The complex physical aspects of the formation of many of the illustrated structures are outside the scope of the book, but those wishing to go more deeply into the subject should find the Further Reading section helpful.

Geological structures have a wide appeal and fascination to geologists and amateur enthusiasts. They are around us wherever rocks are exposed, and it is hoped that this small book will create an interest in, and understanding of, this aspect of geology.

Jointed granite on the Cornish coast.

IGNEOUS ROCKS AND THEIR STRUCTURES

Igneous rocks form from molten material (magma), generated deep within the Earth's crust and upper mantle. This high-temperature material (from 600°C/1,110°F to 1,200°C/2,190°F) is a melt containing many metallic elements, including iron, aluminium, magnesium, potassium, calcium and sodium, together with silicon and oxygen. Various volatiles are also present, especially water. In combination with certain other elements, these form the majority of the minerals that make up igneous rocks. Most of these minerals are silicates, composed of one or more metals, silicon and oxygen, and they include feldspars, pyroxenes, amphiboles, olivines and quartz, the last mentioned being an oxide of silicon.

The composition of an igneous rock depends on the type of original magma, which in turn is related to its source area in the Earth's crust and upper mantle. In the deep continental crust, igneous rocks rich in quartz and feldspar are found, whereas in the ocean areas, igneous rocks tend to be made of less quartz, with more feldspar and ferromagnesian (iron- and magnesium-rich) minerals.

Igneous rocks are classified using two main criteria: composition and grain size.

COMPOSITION OF IGNEOUS ROCKS

According to their mineral content, igneous rocks are separated into a number of groups. The terms 'acid' and 'basic' are used to describe their mineral or chemical composition. Strictly speaking, these words are not completely accurate in the chemical sense, and are based on an outdated theory that silica and water together form 'silicic acids'. The acid igneous rocks include those with more than 65 per cent total silica content. These rocks tend to be pale-coloured, and contain over 10 per cent quartz, often 20 to 30 per cent. Feldspar, especially the variety called orthoclase, is abundant, as is mica. Granite is a well-known example of an acid igneous rock.

Basic igneous rocks are dark in colour, and contain 45 to 55 per cent silica. They are rich in minerals such as plagioclase feldspars and pyroxenes, with only small amounts of quartz. Gabbro is an example of a basic igneous rock.

Between these two groups are the intermediate igneous rocks, which have between 55 and 65 per cent total silica, and contain minerals such as feldspars, pyroxenes, amphiboles, micas and quartz. These rocks have a medium colour and are often speckled in appearance. Syenite is classified in this category.

Rocks with less than 45 per cent total silica content are categorised as the ultrabasic igneous rocks. They are dark, often greenish in colour, and contain virtually no quartz, being composed essentially of pyroxenes, olivines, garnets and amphibole minerals. Peridotite is a typical ultrabasic igneous rock.

GRAIN SIZE AND TEXTURE OF IGNEOUS ROCKS

The size of the grains, or crystals, of which an igneous rock is composed depends on the rate at which the underground magma or subaerial lava has cooled. Slow cooling allows large crystals to develop, whereas if the cooling is rapid, crystals forming the rock will be small.

A rock such as granite, which occurs in huge underground masses called plutons or batholiths, has large crystals, and is referred to as a coarse-grained rock. This coarse-grained category includes rocks with grains or crystals larger than 5mm (0.2in) across. Pegmatite, an extremely coarse-grained, often granitic rock, may have crystals measured in centimetres. The crystals in medium-grained igneous rocks are between 0.5 and 5mm (0.02 and 0.2in) in size. These rocks tend to form in smaller underground masses, including dykes and sills. A typical example is the basic rock dolerite. The fine-grained rocks are those formed on the Earth's surface by the cooling of lava. Here, crystallisation is very rapid, giving the crystals in the rock little time to develop. Basalt and rhyolite are lavas with fine grain size.

The texture of an igneous rock describes the size and shape of its grains or crystals and the relationship between them. A rock is said, for example, to have a coarse-grained equigranular texture if, like many granites, it has equal-sized grains over 5mm (0.2in) in diameter.

A common texture in many igneous rocks is the porphyritic texture, with large crystals set in a finer-grained matrix (groundmass). These larger crystals are called phenocrysts. This texture can be brought about by two stages in the cooling sequence of the magma, which may first cool relatively slowly, producing large crystals. The magma may move higher into the crust, possibly being erupted as lava, and then cools rapidly, surrounding the earlier formed crystals with a finer-grained matrix. Phenocrysts are often of good crystal shape, as they have developed in a relatively liquid material, without being much confined. Such crystals are said to be euhedral. Poorly formed crystals are termed anhedral. It often happens that large crystals in an igneous rock contain and surround smaller crystals. This is called a poikilitic texture.

Diorite and granite, Guernsey, Channel Islands.

▲ These large exposures of two intrusive igneous rocks show differences and similarities. Both of the rocks exhibit a series of cooling joint structures, which have been exploited by marine erosion. The acid granite in the distance across the bay, at Port Soif, is pale-coloured and pinkish because of its orthoclase feldspar content. The diorite, in the right foreground, is typically grey in colour.

When an igneous rock cools very rapidly, its crystals may not even be formed sufficiently well to be seen microscopically. Such rocks have a glassy texture, as in pitchstone and obsidian. Many volcanic rocks are the result of explosive volcanic activity, and these are grouped under the term 'pyroclastic rocks'. The fragmental deposits from volcanoes include tuff, pumice, agglomerate and volcanic bombs. Some of the finer-grained pyroclastic material may be deposited in water, and so exhibit sedimentary structures including stratification, cross-bedding and graded bedding.

STRUCTURES OF IGNEOUS ROCKS

The structures associated with igneous rocks are both those in which the rocks form, including the various types of intrusions, and the structures within the rocks themselves, such as jointing.

As has been discussed, there is a definite distinction between those igneous rocks that form underground and those that cool more rapidly on the Earth's surface. Magma has a tendency to rise towards the upper levels of the crust, and will cool as it rises. Many of the structural features of igneous rocks are a direct result of the way in which the magma or lava has cooled.

Igneous structures include intrusions such as dykes, sills and plutons. Lava flows have a number of cooling-related structures, including surface flow banding and vesicles, which are solidified gas-bubble cavities derived from the volatiles in the magma escaping at higher levels in the crust, especially from lava on the surface. The direction of movement of a lava flow can be determined by the alignment and elongation of the vesicles in the rock. As lava and magma cool, they tend to shrink, and so joints (cracks) develop in the rock. In large intrusions, these may follow a number of directions. When such intrusions are relieved, by erosion, of the weight of overlying rock, new sets of joints may develop in response. In smaller intrusions such as sills and dykes, and in lava flows, a distinct columnar jointing often develops at right angles to the cooling surfaces.

Many igneous rocks are more resistant to erosion and weathering than the surrounding country rock (the rock in which an intrusion formed). For this reason, dykes and sills often give rise to escarpments and ridges, with associated features such as waterfalls. Granite intrusions are noted for their tors, hilltop features resulting from a long period of chemical weathering concentrated on the joint systems in the granite.

INTRUSIVE STRUCTURES

Minor intrusions

Dykes

A dyke is an igneous structure that cuts through existing rocks and their features, such as bedding planes in sedimentary rock, and so is said to be discordant. It is a relatively narrow structure, a few centimetres to a few metres in width, which is vertical or very nearly so, and the magma (molten rock) from which it formed was intruded at no great depth in the crust. The term hypabyssal is used for such relatively shallow intrusions. Magma is intruded from below, generally into a pre-existing fracture, such as a joint or fault, in the Earth's crust. Forceful intrusion may occur, causing these fractures to become widened. The contact between a dyke and the surrounding country rock is often very sharp.

In some areas, including the Inner Hebrides of western Scotland, dykes of Tertiary age occur in swarms, where up to several hundred have been intruded, aligned in the same direction. This suggests that considerable stretching of the Earth's crust has occurred to account for the great volume of magma that has been intruded. In Cyprus, there are numerous dykes that have been intruded into oceanic ridge rocks, now forming the Troodos Mountains in the centre of the island. Research into the rocks here has been important in the theory of sea-floor spreading. Calculations suggest that the crust was stretched about 130km (81 miles) when these dykes were formed. Some dykes can be traced for a considerable distance. The Cleveland Dyke, originating as one of the Hebridean swarm centred on the Isle of Mull, can be mapped as far away as North Yorkshire, a distance of 430km (269 miles).

Dolerite, a medium-grained, dark-coloured rock, is probably the commonest dyke-forming material. It has a basic composition, being rich in dark ferromagnesian minerals such as augite. This rock also contains a high percentage of plagioclase feldspar and up to 10 per cent quartz. Dykes may be composed of various other igneous rocks, including granite, quartz porphyry (microgranite) and diorite. The exact composition is related to the origin of the magma and its cooling history.

Dykes can be intruded into any pre-formed rocks, and can therefore be used as time markers, as they are relatively younger than the surrounding country rocks. As a dyke is made of igneous rock, it is usually possible to date it by radiometric dating methods, which give an absolute age for the dyke rock.

In the field, a dyke is relatively easy to recognise. If seen in a cliff or road cutting, it will be a narrow vertical column of rock, cutting across the local rocks. When on the ground, a dyke is a narrow band of rock, often forming a low ridge, running off into the distance. However, the opposite may be the case, and some dykes are weathered to make vertical, hollowed clefts or sea caves. Detailed examination of a dyke will often show finer-grained rock along its margins, where the magma has been chilled on contact with the colder country rock. Also, the country rock may be metamorphosed (altered) immediately next to the dyke. As the magma cools, columnar jointing can develop, the columns being horizontal in a vertical dyke, as they form at right angles to the cooling surfaces. This columnar structure is often weathered around the margins of the columns. As weathering develops, the angular edges of the columns become rounded.

▶ The sharp contact between the dyke rock, in this case basaltic andesite, and Jurassic sedimentary rocks is clearly seen, showing the different structures of the dipping bedded sediments and the vertical dyke. The andesite of the dyke has no bedding structures, but there is a hint of columnar jointing running almost horizontally through the dyke rock. Otherwise it is massive (structureless). This spectacular vertical sheet of rock is part of the supporting wall of a quarry in the Cleveland Dyke. Originally, the central part of the dyke was quarried for roadstone, and the edges of the dyke were left in place to support the quarry sides. The dyke pillar is about 10m (33ft) tall. It is one of a swarm of dykes that are focussed on the Inner Hebrides, especially the Isle of Mull, where, during the Tertiary era, much volcanic activity occurred and fractures in the crust, radiating some distance, were created.

Dyke, North Yorkshire.

Dyke, Northumberland.

▲ On the coast of Northumberland, at Cullernose Point, there are a number of dykes associated with the intrusion of the Whin Sill. This period of igneous activity is radiometrically dated to 295 million years ago. Here, a small dyke of dolerite forms a thin ridge across the wave-cut platform. This results from the dolerite being harder than the country rock (limestone and sandstone) and more resistant to erosion by the sea. This differential erosion is typical of landform development associated with igneous rocks. The narrow ridge is around 2m (6.6ft) high.

► (top) The second picture, taken at Cullernose Point, shows the contact of the dolerite dyke with bedded limestone against it. The limestone country rock is stratified, whereas the dolerite is massive and blocky, showing the very striking differences between sedimentary and igneous rocks. The limpet shells are about 3cm (1.2in) in diameter.

► In contrast with the upstanding dyke in Northumberland, these dolerite dykes on the Cornish coast, at Pentire Point, have been eroded more readily than the surrounding slate, producing vertical clefts in the cliff. As well as marine erosion, water running from the cliff top has also been active in their removal. The dykes are about 3m (9.8ft) high.

Contact of dyke and limestone, Northumberland.

Eroded dykes, Cornwall.

Eroded dyke, Guernsey, Channel Islands.

▲ Landforms associated with eroded dykes include caves, as here on the east coast of Guernsey, at Marble Bay, where gneiss has proved much more resistant to erosion than this dolerite dyke. The cave is about 2.5m (8.2ft) high.

▶ Where a dyke is in contact with the colder country rock, it often has a chilled margin. This pinkish granite was formed around 650 million years ago, and the dolerite dyke was intruded 296 million years before the present. The granite was therefore relatively cold, and chilled the dyke magma rapidly along its margins. This picture shows the granite in detail, with its abundance of pink feldspar and grey quartz, and the sharp contact. Right next to the granite, the dolerite of the dyke is very fine-grained where the margin is chilled, but away from the contact its grain size increases.

Dolerite dyke, Guernsey, Channel Islands.

▲ On this low cliff, at L'Éree, Guernsey, a grey dolerite dyke is seen, cutting through pale pinkish granitic country rock. The dyke is 2m (6.6ft) wide. These igneous rocks clearly illustrate the difference between dark basic rock and pale acid rock. The pink granite is speckled with coarse crystals, but crystals in the medium-grained dolerite are much smaller. The dolerite magma has been intruded into a major joint in the granite. Various fractures are seen in the pinkish granite, but the main joints in the dyke tend to run across the intrusion, at right angles to its vertical cooling surfaces.

Chilled margin of dyke, Guernsey, Channel Islands.

Dyke swarm, Guernsey, Channel Islands.

▲ Dykes often occur in large groups called swarms, and these three dykes in the southern cliffs of Guernsey, at Icart, are part of a group of many basic dykes that are exposed along this coast. A swarm of dykes may occupy a series of fractures in the crust, which usually have the same orientation, and are related to an episode of tectonic activity. The country rock here is gneiss. The dykes in the picture have widths of about 1 to 3m (3.3 to 9.8ft).

▶ This coastal exposure (top), at Achmelvich, Sutherland, shows the margin of a dark basic dyke, about 20m (66ft) wide. The dyke is intruded into paler, highly metamorphosed Lewisian gneiss. A large, angular fragment of gneiss forms a xenolith (inclusion) within the dyke, and is seen near the top right of the picture. The altered margin of the dyke has been more readily eroded, creating a deep cleft.

▶ Though basic rocks such as dolerite are common in dykes and sills, other rocks also form minor intrusions. This narrow dyke, cutting through diorite country rock, is composed of pale pinkish granite, contrasting with the darker diorite. Joints in the dyke have developed horizontally, at right angles to its cooling surfaces.

Dyke, Sutherland.

Granitic dyke, Guernsey, Channel Islands.

Sills

A sill is a concordant igneous intrusion, in contrast to a dyke, which is discordant. The word concordant indicates that the magma forming a sill was intruded between existing rock structures, such as strata, rather than across them. Unlike a dyke, a sill is often horizontal, or nearly so, though it may move upwards and transgress from one stratigraphic level to another. It may be fed with magma from below by a magma pipe, which on solidifying becomes a dyke.

With a dyke, it is relatively easy to understand how the magma rises up into an existing fracture, but in the case of a sill, it is more difficult to appreciate how the magma is intruded, when there may be a considerable thickness of rock above it. The pressure at which the magma is intruded must be greater than pressures caused by the weight of overlying strata. For this reason, it is thought that a sill is usually intruded at a relatively shallow depth, and may be injected into unconsolidated sedimentary rocks.

Dolerite is a typical sill-forming rock, but sills may also be made of a wide variety of igneous rocks. Columnar jointing is a common structure in sills. It develops vertically through the intrusion at right angles to the cooling surfaces at the top and bottom of the sill.

Sills are often of no great thickness, but some may be over 30m (98ft) thick. The Palisade Sill, which is exposed along the Hudson River in New York, reaches a thickness of about 300m (980ft).

Because the rock of which a sill is made is very often far more resistant to erosion and weathering than the local country rocks, sills can give rise to landforms including escarpments and linear ridges. Where rivers encounter such resistant rock, waterfalls are common.

▶ This dolerite sill, at Salisbury Crags, Edinburgh, was intruded into the side of the Arthur's Seat Volcano about 25 million years after the end of the volcanic activity of 340 million years ago. The vertical jointing in the sill is a typical structure of such an igneous intrusion. Here, the thickness of the sill is around 43m (140ft). The magma was injected among sedimentary rocks, and these have suffered contact metamorphism, heat from the sill-forming magma having altered the sediments adjacent to the base and top of the sill.

The closer picture shows how the dipping sedimentary rocks below the sill have been whitened by heat from the magma. Sills have many

Sill, Edinburgh.

features that are similar to lava flows, and it can be difficult to distinguish between them in field exposures. If the overlying rocks have not been removed by erosion and are metamorphosed, then the igneous body is a sill, as a lava flow can only affect rocks over which it flows. The dipping sedimentary rocks, exposed below the sill, show stratification, in contrast with the vertical jointing of the sill. As with many sills, Salisbury Crags forms a steep escarpment, here with the dip slope to the right.

Transgressive sill, Edinburgh.

▲ The rusty-coloured base of the Salisbury Crags sill can be seen disrupting a block of sedimentary country rock and rising above it. Here, the sill transgresses to a higher level, though for most of its lower contact it remains at the same horizon. These underlying sedimentary rocks have been whitened by the metamorphic effect of the intruded magma. This exposure is referred to as 'Hutton's Section' and is a world-famous location. James Hutton (1726–1797) has been called 'the father of modern geology'. He was a Scottish geologist whose ideas, based on field observations, suggested that certain rocks such as granite were formed in a molten state, and that geological processes we can observe today have been taking place throughout geological time. Hutton visited Salisbury Crags and studied the lower contact of the sill and sedimentary rocks, noting the way the sill transgresses from one level to a higher one.

▶ The Whin Sill, which is exposed through Durham and Northumberland, is a concordant dolerite intrusion, dated at 295 million years. It intrudes rocks of Carboniferous age, and in places, especially Upper Teesdale, has an important metamorphic effect on the country rocks. In this area, the limestone of Lower Carboniferous age has been altered into a 'sugar limestone', with a granular texture. The sill has typical vertical jointing, which distinguishes it easily from the bedded country rocks. In certain areas, especially on the west-facing escarpment of the Pennines near Appleby, the Whin Sill transgresses from one horizon to another.

The word whin, or whinstone, is an old quarrying term for the resistant, dark-coloured dolerite, that has been used for hundreds of years as a durable roadstone. The sill has considerable impact on the landscape, forming various landform features, mainly because of its slight dip and its resistance to erosion, when compared with local country rocks.

As well as an escarpment, exploited by the Roman army in the building of Hadrian's Wall from AD 122, this sill forms many waterfalls and offshore islands, including the Farnes, renowned for their nesting seabird colonies. High Force Waterfall in North East England is a landform caused by the hardness of the dolerite of which the Whin Sill is made. The picture, taken from the southern side of the River Tees, shows the vertically jointed igneous rock over which the river falls. Beneath this dolerite are bedded layers of limestone and sandstone of Carboniferous age. These sedimentary rocks are more easily eroded by the river, and so a gorge has been created downstream from the waterfall. The fall here is 21m (69ft) high.

The Whin Sill, High Force Waterfall, County Durham.

Whin Sill, Northumberland.

▲ Here, the Whin Sill has been used as a strategic position since before Norman times. The sill can be seen beneath the castle, at Bamburgh, with Carboniferous sandstones below it. Across the left-hand half of the picture, the base of the sill transgresses from one stratigraphic level to higher ones. There is a striking difference between the dark grey, blocky dolerite of the sill and the pale-coloured sandstones below.

▼ The hardness of the dolerite sill has produced a steep sea cliff and headland. The jointing in the sill is mainly vertical, but at the seaward end of the headland it tends to be slightly inclined. In the extreme foreground are Carboniferous sedimentary rocks, into which the sill has been intruded.

Whin Sill, Northumberland.

Transgressive sill contact, Northumberland.

▲ The base of the Whin Sill rises from right to left across this picture (top), transgressing over Carboniferous limestone and shale beds. The difference between the structure of the blocky igneous dolerite and the bedded sedimentary limestone is well shown. The height of this low cliff is around 4m (13ft).

The lower picture shows the Carboniferous limestone surface, with the sill resting on it. The sill has transgressed up from a lower level on the far right.

Xenolith in dolerite sill, Northumberland.

▲ The irregular area within the sill dolerite is a xenolith (inclusion) of altered limestone. This has been taken in by the molten sill magma at the time of intrusion. Xenoliths are generally taken as evidence for forceful intrusion of magma. This exposure is on the coast of Northumberland.

Flow structure in dolerite sill, Northumberland.

▲ As is often seen in magma that has been mobile, the dolerite magma of this sill shows small sections of flow structure with a ropy appearance. The direction of flow of the molten rock can be worked out from structures such as this and careful mapping.

Major intrusions

Large-scale igneous intrusions form as structures called batholiths and plutons, vast discordant bodies of rock, usually composed of granite, though syenite and gabbro also occur is such structures. These intrusions can be many tens of kilometres across, and frequently occur in mountain regions. The large batholith in South West England is 65km by 40km (40 miles by 25 miles) in area and has upward projections called cupolas, which, as a result of millions of years of erosion of overlying strata, are now exposed in high moorland areas. In Peru and the western coastal mountains of Canada, there are batholiths over 1,000km (620 miles) long. It is probable that batholiths are formed by several sequences of intrusive activity, and a large batholith may take many tens of millions of years to consolidate completely.

Batholiths vary in their overall shape. Some have steeply sloping outer margins, while others are more uneven in structure and may be more elongated. Batholiths form at considerable depth; the upper margins are possibly 10km (6.2 miles) below the surface when intruded. Outlying masses that are usually joined to the main body of igneous rock are called stocks.

It is not certain how such large masses of magma are injected at depth in the Earth's crust. A batholith or pluton has a great volume and occupies a large space within the crust. This 'space problem' has been considered for many years. There is evidence to suggest that, at least in part, magma can melt its way into the crust from below and assimilate some of the surrounding country rock. Xenoliths are frequently found within the igneous rock around the margins of large igneous intrusions. These are blocks of country rock that have become incorporated in the magma and altered. They often occur as dark-coloured, rounded masses, commonly of no great size, which may contain feldspar and other crystals typical of the surrounding igneous rock. These xenoliths are taken as direct evidence of the magma partially melting its way into the country rock.

A process that accounts for xenolith formation is called stoping. This suggests that the intruding magma rises though weaknesses in the country rock, surrounding and assimilating masses of it. Evidence also implies that some large igneous intrusions are forcefully created. There is often a sharp contact along the intrusion's margins and buckling of the country rock. Salt domes, such as those in southern Iran, show forceful intrusion where they have risen through the crust. This has some similarities with the way in which large igneous masses are intruded.

As the granitic magma rises up in the crust and cools, large crystals are able to form and volatiles (materials that at normal temperatures would be gases) in the magma may crystallise to form rare minerals. The usual minerals of granite (feldspar, mica and quartz) commonly produce a coarse-grained equigranular texture, but with different cooling stages, a porphyritic texture can form. This has large crystals set in a finer groundmass. The magma has a profound effect on the surrounding country rocks, causing them to be altered by contact metamorphism, often with recrystallisation of the country rock.

The rocks in large intrusions typically have joint structures running through them. These joints commonly occur both vertically and horizontally and may be a direct result of shrinkage during cooling. Some joints, sometimes referred to as unloading joints, are formed by stretching, as the great thickness of overlying country rock is removed by erosion. Joint systems are important in the formation of granite landforms, allowing acid rain and groundwater to enter and chemically alter minerals in the granite, especially feldspar, which is changed to clay minerals. These are then washed out and the joints become enlarged. Hydrothermal fluids, generated at depth, can rise through the joint system in the granite. These highly active fluids can affect the feldspar in the rock even more profoundly than acid rainwater. The unaltered granite cores may form tors when exposed.

▶ A small (20cm/7.9in) mass of volcanic rock has become incorporated in paler granite during emplacement of magma. This xenolith has been so altered that pinkish feldspar crystals have formed within it. The photograph shows a boulder of granite from Shap in Cumbria on the shore at Runswick Bay, North Yorkshire. Along with many others, this is a glacial erratic, indicating ice movement from the west during the last glaciation.

Granite contact, Cornwall.

▲ The contact regions of large igneous intrusions can be sharp, as in this exposure, or more complex and irregular. Here, the contact between pinkish granite, dipping steeply at around 30°, and dark slate is seen. The slate, originally shale, of Devonian age, was regionally metamorphosed before the intrusion of the granite around 290 million years ago. These slates have been thermally metamorphosed by the granitic magma, and are now flinty hornfels next to the contact with the granite. Xenoliths containing black tourmaline occur within the intrusion.

Xenolith in granite.

Xenoliths in jointed diorite, Herm, Channel Islands.

▲ On the small island of Herm, an intrusion of granodiorite contains numerous rounded xenoliths of darker rock. These xenoliths are mainly composed of diorite, another igneous rock. Their presence here has caused much debate. It is generally believed that these xenoliths come from a large mass of diorite, which was surrounded by the granodiorite magma during its intrusion, partly broken up and incorporated in the granodiorite. The largest xenoliths are around 30cm (12in) long.

▶ (top) Rocks forming large discordant intrusions, especially granite, characteristically exhibit jointing structures. Here, the reddish-coloured granite on the south-west coast of the Isle of Mull, Scotland, looking across to Iona, shows vertical joints, which have been enlarged by weathering and erosion to produce a series of rough vertical blocks. The pink feldspar and pale quartz crystals in the granite are visible.

▶ Vertical joint structures in this exposure of granite have been exploited by marine erosion and this has led to the formation of a coastal arch about 10m (33ft) high. Eventually the roof will collapse, and a sea stack will remain.

(above) Jointing in granite, Isle of Mull. (below) Jointing in granite, Cornwall.

Rough Tor, Cornwall.

▲ Like many granite tors found in South West England, Rough Tor on Bodmin Moor shows the joint structures typical of this rock. Here, there are both horizontal and vertical joints. The main tor here is about 10m (33ft) tall.

Tor formation has been debated for many years. There is certainly a strong link between the granite joint structure and the landform. Other rocks with joint structures, including sandstone, can also form tors. Essentially, a tor is a rocky summit landform that rises from smoother hill slopes. It may be that deep rotting of the granite along its joints caused by acid waters produces enlarged joints. Feldspar in the granite is especially susceptible to chemical weathering. When the overlying regolith (loose rock and soil) was removed, as happened during the late stages of the Quaternary glaciation, the granite cores remained.

▶ (top) These mountains of Cretaceous granite, at Yosemite, California, have been extensively glaciated, producing deep, steep-sided valleys. The vertical joint structures that can be seen in the cliffs are probably cooling joints. The famous Half Dome is visible in the centre of the picture.

Jointing in granite, California.

▶ A granite exposure on the south coast of Cornwall, at Porthcurno, shows joint structures and large feldspar phenocrysts in a finer matrix. The phenocrysts are about 5cm (2in) in length and aligned in the same general direction. This suggests that they were formed in relatively molten magma, which did not confine their growth, and the subsequent movement of the magma, together with their tabular shape, brought them into subparallel orientation, as a flow structure.

Porphyritic granite, Cornwall.

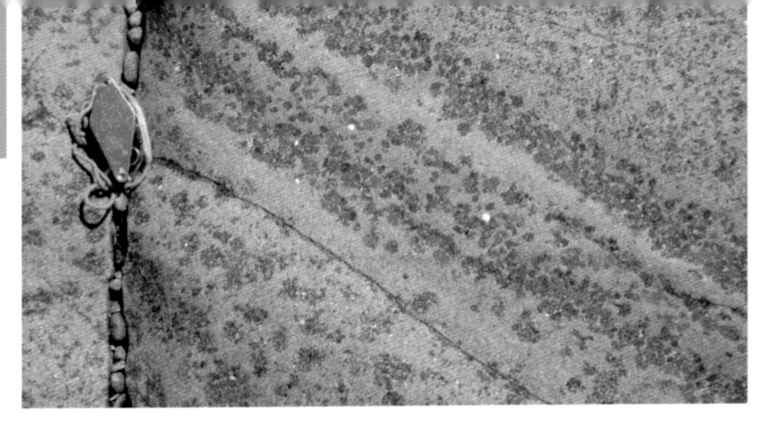

Layered gabbro, Guernsey, Channel Islands.

▲ It is not uncommon to find layering structures in the coarse-grained igneous rock gabbro. This rock has a basic composition, made of plagioclase feldspar, pyroxene minerals such as augite, and amphiboles including hornblende, which gives the rock an overall dark colour when compared with the much paler acid rock, granite. In this example of layered, or banded, gabbro from eastern Guernsey, the darker bands are rich in hornblende while the paler grey matrix of the rock is predominantly plagioclase feldspar.

The origin of these bands is open to debate, but it may well be related to the cooling of the parent magma and the order of formation of the crystals making up the rock. The dark, relatively dense mineral crystals, such as the hornblende seen here, will cool early, especially around the walls and roof of the intrusion, sinking to form a layer rich in these minerals. Subsequently, paler, feldspar-rich rock crystallises. This theory may well explain the production of a single set of layers in an intrusion, but doesn't fully account for repeated layering. Various suggestions have been made to account for this structure, including the effects of convection currents within the magma and the repeated injection of fresh magma into the intrusion, but none of these is wholly satisfactory. It seems that different processes may be at work in different intrusions.

▶ (top) This layered structure in an intrusion of coarse-grained diorite, an igneous rock of intermediate composition, is similar to that found in gabbro. The origins may be the same, and the banded structure could result from gravity settling of early formed crystals within the molten magma, causing segregation into layers. However, it has been suggested

Layered diorite, Guernsey, Channel Islands.

that alteration by metasomatism (the action of hot fluids) may also have contributed by reworking the original diorite. This section, at Beaucette, is about 3m (9.8ft) in height.

▼ This structure within the main mass of gabbro shows small, dark, angular fragments of hornblende-rich rock, surrounded by a network of paler, more acidic veins made of a rock called aplite. In the darker gabbro fragments, thin prismatic hornblende crystals can be seen. A theory that may explain the structure seen here suggests that after the consolidation of the gabbro, a more acid magma was intruded under pressure. This found its way into small fractures in the rock, surrounding and breaking fragments off the gabbro. It has been suggested that this more acidic magma could be a residual material, left after the bulk of the gabbro-forming magma crystallised, and now depleted of the minerals necessary to produce a dense basic rock.

Fragmented gabbro, Guernsey, Channel Islands.

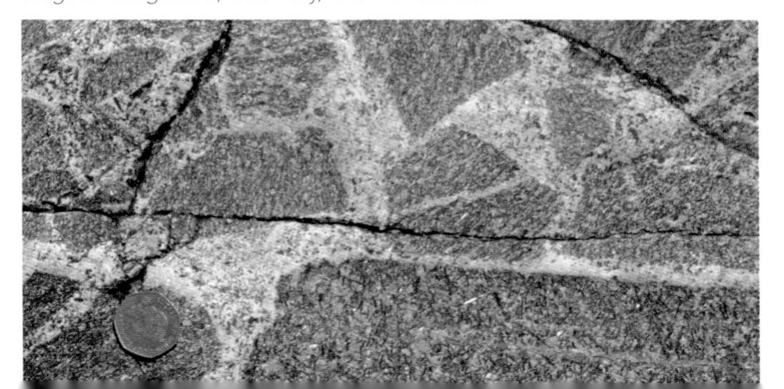

Tourmaline vein, Cornwall.

▲ Large-scale igneous structures, including intrusions such as batholiths, produce considerable hydrothermal activity. This is mainly associated with high-temperature waters, rich in the chemical components of many minerals. This thin vein, which occupies a joint in granite, is filled with black tourmaline, a characteristic mineral found in veins associated with granite. The hydrothermal activity may not have taken place at the same time as the intrusion of the igneous structure, but can happen many millions of years after the initial intrusion.

◀ Pale quartz is a common mineral in veins related to granitic structures. These irregular veins are in slate near the margin of a granite mass in south Cornwall. Quartz veins are often associated with other minerals. These are frequently metallic, and can contain important ores such as galena (lead sulphide), sphalerite (zinc sulphide), cassiterite (tin oxide) and chalcopyrite (copper iron sulphide).

Quartz vein, Cornwall.

STRUCTURES OF EXTRUSIVE IGNEOUS ROCKS

A volcanic eruption produces many different rock materials, including ash, rock fragments, various gases and water. These pyroclastics are usually blown from the volcano first, followed by lava, which requires the lowest pressure during the eruption.

The structures formed by lava flows and other volcanic material depend mainly on the type of lava erupted. Acidic lavas tend to be viscous (sticky), and therefore do not flow very far. These are derived from the type of magma that would form granitic rocks if solidified at depth. Such lavas are rich in silica, and usually have a low gas content. When ejected, they can give rise to explosive eruptions. These can be a result of the lava solidifying within the volcanic vent and needing a considerable build-up of pressure before the next eruption can occur. When this happens, blocks of lava and rock from the sides of the volcano may be erupted.

Basic lavas tend to be at a much higher temperature and have a greater gas content, giving them a lower viscosity. These lavas flow over wide areas, though the individual flows may be relatively thin, only reaching a maximum thickness of about 10m (33ft). Lava flows may exhibit features such as columnar jointing and ropy surface structures. Two Hawaiian words are used to describe different types of lava: aa is the name given to blocky lava, while pahoehoe has a smoother, ropy surface.

Pyroclastic deposits are those formed by the ash and fragmental material generated by a volcano's explosive activity, which may settle on land or beneath water. Thinner deposits with smaller fragments are found farther from the volcanic vent. The finest ash particles are carried by prevailing winds, often great distances from the volcano. Such deposits can be of considerable stratigraphic value as time markers, having settled over a wide area at the same time. Ash fragments that are deposited in water may have many of the features of sedimentary rocks, including graded and cross-bedded structures. Larger particles (over 65mm/2.6in diameter) settle near to the vent or even within it, and form a rock called agglomerate.

Probably the most violent and devastating pyroclastic eruptions are those of acid lava volcanoes. The incandescent pyroclastic flow called a nuée ardente eruption is a cloud of molten lava droplets and gas, which can move downhill at over 60kph (40mph). The deposit from this flow can fill hollows and valleys in the landscape, and settle to form a welded tuff called ignimbrite. This and many other lava flows exhibit a type of layering structure related to the movement of the lava. This is known as flow banding. If pyroclastic material is erupted during times of heavy rainfall, volcanic mudflows may result.

Lava flows can usually be distinguished from concordant structures such as sills by a number of features. A sill, because it is intrusive, will have a chilled margin at both its upper and lower surfaces. The rocks above and below a sill may show contact metamorphism. A lava flow can only have these features below it, and may have a weathered upper surface. The vesicular and amygdaloidal textures of many lavas, especially those of basic composition, are generally in the upper parts of a lava flow, as the gas in the molten lava tends to rise towards the surface.

A number of fine- and medium-grained igneous rocks cool in such a way that columnar joint structures develop. The columns are often exactly hexagonal in cross-section, but can vary and be more random in outline. This columnar jointing is related to the structure of the igneous body. Contraction of magma takes place initially along its cooling surfaces, and here the pattern of jointing first occurs. As the mass of the magma continues to solidify, so the joint pattern develops through the body of the rock. In a lava flow, the columnar structure tends to be vertical, as the flow cools at first along its base and top.

Minor intrusions can also exhibit columnar jointing. In a dyke, because the intrusive structure is vertical or nearly so, the columns usually develop horizontally, at right angles to the cooling surfaces. Columnar jointing in a sill is usually vertical.

▶ (top and centre) Almost perfect hexagonal columns have formed in basalt of Tertiary age on the Island of Staffa, Inner Hebrides, Scotland. Very regular cooling, extending through the lava flow, would have been necessary to develop such well-formed columns. Fingal's Cave, near the centre of the picture, is about 20m (66ft) high.

▶ Columnar jointing is most common in fine-grained igneous rocks, as here, in basaltic lava, where the columns are many metres high, with a consistent cross-section.

Columnar jointing, Isle of Staffa, Scotland.

(below) Columnar jointing, Samson's Ribs, Edinburgh.

Columnar jointing, Hvita Gorge, Iceland.

▲ In this exposure, a basaltic lava flow exhibits columnar jointing and forms a steep, valley-side cliff. Below the flow are bedded layers of volcanic ash.

▶ (top) Craters are typical structural features of volcanoes. Here at Kerið, in southern Iceland, the volcanic crater is occupied by a lake. Several crater lakes occur in this region, which was volcanically active some 3,000 years ago. Layers of basaltic lava can be seen in the crater walls to the left of the picture. The crater is 55m (180ft) deep and 270m (890ft) across, but the lake, with its amazing blue colouring due to minerals from the surrounding rocks, is only up to 14m (42ft) deep. The lake is maintained by the level of underground water (the water table) in the region. Many such crater structures are the result of explosive volcanic activity. However, there is no geological evidence for such an explosion at Kerið. It is thought that after erupting and depleting its underground magma chamber, the volcanic cone collapsed, creating the deep crater.

▶ This tuff, formed from a nuée ardente eruption, shows typical flow structures and numerous small shards welded together. These can be seen running across the specimen, which is from the Ordovician volcanic rocks of North Wales. The specimen measures 10cm (3.9in) across.

Volcanic crater Iceland.

Ignimbrite, (welded tuff), North Wales.

Volcanic crater of Poas, Costa Rica.

▲ This stratified volcanic crater contains a lake of such acidic water that at times the pH reads virtually 0. The volcano reaches a height of 2,700m (8,800ft) above sea level. The beds of ash and tuff can be seen in the crater walls, with structures very like those of sedimentary strata. The lava is mainly intermediate in composition, andesite being typical.

► (top) Here, an inland cliff shows a lower layer of bedded volcanic ash, which is finely stratified like a sedimentary rock. Above it, with a sharp contact, is dark basaltic lava, which has a blocky structure and vertical joints. The ash has weathered into a fine, dusty scree, and the lava into large blocks, broken off along its joint structures. The exposure is about 3m (9.8ft) high.

► An inland cliff exposure shows thin, interbedded lava flows and red-weathered ash deposits with stratification structures. The agglomerate layers contain both fine-grained ash and larger volcanic fragments, indicating that the deposit has formed only a short distance from the volcanic vent. The height of the cliff is about 10m (33ft).

Lava flow on bedded ash, Iceland.

Stratified lava and agglomerate, Iceland.

Red bole, Isle of Skye.

◀ Lava can be weathered soon after its eruption, particularly by chemical processes. This is especially the case when there is a time gap between eruptions of successive lava flows. Here, basic lavas of Tertiary age have reddened upper surfaces due to chemical weathering. The red colouration is due to oxidation of iron-containing minerals, producing haematite, an iron oxide. As this colouring is on the upper surface of the lava flow, it is a feature of stratigraphic value, which can be used to determine the 'way-up' of the rocks. This exposure is about 6m (20ft) in height.

◀ Waterfalls are landforms often controlled by rock structure. Here, a resistant basalt lava flow forms an inland cliff, too resistant for easy river erosion. This is part of a cliff line that, in the recent past, about 3,000 years ago, was on the Icelandic coast. The sea level has fallen and the cliffs are now 5km (3 miles) inland. The Skógafoss waterfall is 60m (200ft) high.

Skógafoss, Iceland.

Ropy lava, Iceland.

▲ ▶ This type of lava is often known by the Hawaiian name 'pahoehoe'. It is formed when the lava surface, on contact with the cool atmosphere, begins to solidify. The molten lava beneath continues to move and flow, creating corrugations in the lava's cooling surface. Basaltic lavas that flow quickly often develop this structure, which is well known in areas of shield volcanoes. Frequently, as in the picture from Hawaii, ropy lava forms on slopes, especially where molten rock moves over irregularities.

Ropy lava, Hawaii.

Bedded agglomerate, Fife.

▲ This volcanic rock of Carboniferous age, in Fife, Scotland, consists of a fine-grained tuff containing larger fragments of volcanic and other rock, probably broken off surrounding and underlying areas. These vary in size from a few centimetres up to about 30cm (12in) in diameter. The large blocks in this deposit suggest that considerable volcanic force has propelled them from the vent. The largest blocks are aligned in a roughly horizontal bedded structure, as are some of the smaller fragments. This bedding may have an inclination and dip related to the original volcanic slopes on which it was deposited. In such deposits, it may, therefore, be possible to reconstruct the size and position of a volcanic crater by using these structures, along with other information obtained from lava flows and related volcanic phenomena.

▶ (top) These rounded structures are pillow lavas, exposed in Anglesey, North Wales, formed by submarine volcanic eruptions. When basaltic lava is erupted under water, a skin of consolidated lava forms very quickly on contact with the cold seawater. The molten lava within this skin may continue to be erupted, filling the pillow and expanding it. This can also happen beneath ice flows, as in Iceland; a subglacial eruption melts the ice forming a meltwater cavity in which pillow lava forms. Pillow structures may have vesicular margins, a radiating joint pattern and a very fine-grained, glassy crust.

In some cases, lava may burst from an already formed pillow structure, allowing a new pillow to form. The pillows tend to build up in irregular masses. Andesitic and rhyolitic lavas, erupting under water or ice, also produce pillow structures, but these are usually many metres across, unlike the smaller basaltic pillows. In the fissures between the individual pillows, a variety of materials may occur. This ranges from chert to sedimentary rock, often shale. The overall assemblage of materials will give an indication of the environment in which the eruption took place. This low cliff is about 4m (13ft) high.

Pillow lavas, Anglesey, North Wales.

▼ This close-up picture of the rounded margins of two lava pillows shows brownish weathered basalt containing numerous small vesicles. These are gas-bubble cavities in the lava. The pillows are separated by a mass of greyish chert. Being silica-rich, this has resisted the weathering that has altered the basalt. Chert is commonly found between pillow lava mounds. It may be formed from the internal skeletons of microscopic marine planktonic organisms, especially radiolarians. This region of the Cornish coastline, at Pentire Point, is a classic site for the study of submarine lavas, including these pillow structures. The area shown is about 50cm (1.6ft) across.

Pillow lava, Cornwall.

Blocky lava, Iceland.

▲ This type of lava is often known by its Hawaiian name of 'aa' lava. It is usually the product of fissure eruptions and has a rough surface structure. As it forms, the mass of lava moves forwards slowly. The steep moving front of the lava flow is generally a mass of clinker and glowing lava. Eventually, the whole mass cools, to make an area of irregular, dark lava blocks of varying sizes, which is very difficult to walk over.

▶ (top) Here, in a cliff exposure some 5m (16ft) high, layers of tuff have been deposited in neat strata with the appearance of sedimentary rock. The upper grey layer is an ignimbrite flow. It was created from silica-rich, viscous lava that was filled with gas. The yellowish, lower layer is a clastic air-fall deposit associated with a violent eruption. The layers vary from fine-grained ash to coarser tuff containing fragments of lava.

▶ When volcanic products, especially fine-grained ash, are ejected from a crater to be deposited in water, they settle and have the characteristics of a sedimentary rock. Here, a very fine-grained tuff deposit shows the typical features of sedimentary graded bedding, with many repeated units present. It may be that the tuff was reworked by water currents to produce this structure.

Ignimbrite and tuff, Costa Rica.

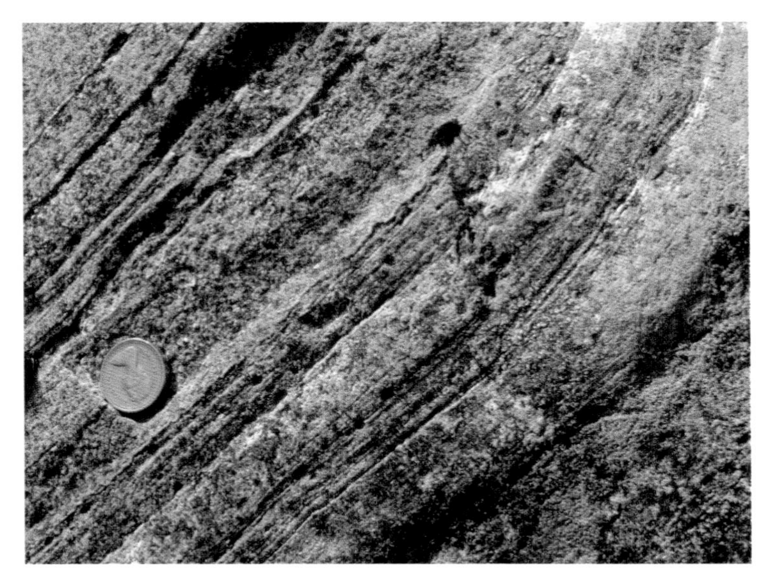

Graded bedding in tuff, Cumbria.

Cross-bedded tuff, Cumbria.

▲ A further example of sedimentary bedding structure in volcanic ash is seen in this cross-bedded Ordovician volcanic rock. Ash erupted high into the atmosphere has fallen into flowing water, to produce this cross-bedded structure.

▶ (top) This small island, bounded by vertical cliffs, is situated off the south-east coast of Scotland. At 110m (360ft) high, it is a resistant volcanic plug, composed of trachyte lava. The tuff of the original volcano has been eroded away since its eruption during the Carboniferous period. Resistant igneous rock frequently creates landforms such as this. The white seabirds are gannets, which nest in a huge colony of over 150,000 birds.

▶ Here, successively erupted basaltic lava flows of Tertiary age form a stepped hillside. Each flow is more resistant to erosion than the intervening volcanic tuff and sediment. Escarpments and stepped features such as these are characteristic landforms associated with resistant igneous rocks. The basalt has vertical joint structures, which have developed at right angles to the cooling surfaces of each lava flow.

Volcanic plug, Bass Rock, Scotland.

Lava flows, Isle of Mull.

METAMORPHIC ROCKS AND THEIR STRUCTURES

Metamorphic rocks are created when pre-existing rocks are subjected to various forces that cause them to change into new rocks. The original rock's texture, mineralogy and structure may be altered. Sedimentary, igneous and pre-formed metamorphic rocks can all be metamorphosed. Changes of this type occur when temperatures and pressures increase, often at depth in the Earth's crust, though high temperatures from lava erupted on the surface can also metamorphose rocks with which it is in contact. Temperatures below about 200°C (390°F) generally have little or no effect, metamorphic changes usually taking place between 200°C (390°F) and 700°C (1,290°F). Pressure increases dramatically with depth, and at 20km (12 miles) is around 6,000 atmospheres (88,000lb/sq in). This will also increase during mountain building, resulting from plate tectonic activity. If temperatures and pressures are extreme, rocks may be melted and igneous processes initiated.

CONTACT METAMORPHISM

When magma and lava are fluid, they may reach temperatures of about 1,000°C (1,830°F). Any cold crustal rocks with which the magma or lava is in contact (the country rocks) may be altered by contact (thermal) metamorphic processes. The mineralogy of the heated rocks changes and new metamorphic minerals form, both from the original rock's minerals and from hot fluids that may permeate the country rocks.

A large intrusion, such as a batholith, will have a widespread metamorphic influence, possibly measured in hundreds of metres. A small dyke or sill may only metamorphose its country rocks for a few centimetres on either side. The area of metamorphism around an intrusion is called the metamorphic aureole. It has been calculated that the country rock temperature near to an intrusion with a diameter of around 10km (6.2 miles) will fall from around 700°C (1,290°F) to 400°C (750°F) at a distance of 4km (2.5 miles).

A much smaller intrusion, only 1km (0.6 miles) in diameter, will only heat country rocks to 100°C (210°F) at around the same distance.

As well as the size of the igneous body, its composition is critical. Granitic magma often produces much high-temperature, water-based fluid, which invades the country rocks, producing mineralogical changes. The nature of the country rocks under the influence of high temperatures is also important, some rocks resisting metamorphism more than others. Quartzite and sandstone and many igneous rocks are far less easily altered than shale, clay and volcanic ash. Contact metamorphosed rocks tend to lack structural features, but their occurrence indicates the presence of igneous structures and is evidence of the nature of the igneous intrusion.

Rocks subjected to contact metamorphism have a recrystallised texture and their mineralogy is related to the original unmetamorphosed rock, with additional minerals produced by the heating processes. When volcanic tuff and fine-grained sedimentary rocks, including shale, clay and mudstone, are altered by heating, a rock called hornfels is created. This is generally a fine- to medium-grained rock, noted for its tough, often flinty, appearance. Original sedimentary structures such as bedding are usually missing, having been removed by recrystallisation. The rock is rich in quartz, and new metamorphic minerals such as andalusite and cordierite are often present. The rock may have a granoblastic texture, where larger crystals have developed. These may be of garnet or chiastolite.

Marble is a contact metamorphosed rock, produced when limestone is heated. Impurities in the original limestone may be altered into new minerals, such as diopside and brucite, which can give marble distinctive colour bands and veins. Calcite from the limestone is recrystallised, and grows to give marble its distinctive texture.

Sandstone is recrystallised in a similar way to limestone, the original quartz grains fusing and growing to create a rock called metaquartzite. The quartz content may be over 90 per cent of the rock.

Contact metamorphic effects of a dyke, North Yorkshire.

▲ Here, the metamorphic effects of the Cleveland Dyke in North Yorkshire are readily observed where a quarry wall has fallen away. The picture clearly shows how certain rock types resist thermal alteration, while others are metamorphosed, and the structural differences between the sedimentary and igneous rocks are well displayed. The dyke rock, basaltic andesite, is the dark, blocky material at the base of the quarry face. The sedimentary rocks, shale and sandstone, have almost horizontal bedding structures.

When intruded, the igneous rock may have been at a temperature approaching 1,000°C (1,830°F), but different country rocks are affected in different ways by metamorphism. The pale layer immediately above the igneous rock is metamorphosed Jurassic shale, which, because of its clay mineral content and fine-grained, poorly cemented nature, has not resisted heating by the igneous rock and has been considerably altered and whitened. Above the shale are beds of brown-coloured sandstone. This rock, being rich in quartz, has suffered less metamorphism than the shale.

REGIONAL METAMORPHISM

Unlike the fairly local effects of contact metamorphism, regional metamorphism can affect very wide areas. It occurs at some depth in the Earth's crust, where temperatures and pressures are high, when folding of rocks occurs during periods of mountain building. Three main grades of regional metamorphism are recognised: low, medium and high. In each grade, typical metamorphic rocks are found, often with mineral assemblages that can be related to the pressure and temperature conditions that created them. These conditions are controlled by depth of burial and proximity to the intense areas of folding. Regionally metamorphosed rocks have many structural features related to their formation.

There are three structural concepts that are related to rock deformation and are recognised especially in regionally metamorphosed rocks. These are rock fabric, lineation and foliation.

Rock fabric concerns the particles in a rock that have a geometrical orientation related to the stress that a rock has suffered. It has been defined as the rock's 'structural texture', and is a combination of all the features of a rock. The orientated particles in the rock may be aligned as surfaces or as lines running through the rock. Usually rock fabric is seen as the small-scale features of a rock, often those requiring a hand lens to determine details. Fabric is generally the response of the particles of which the rock is composed to deformation, and this structure gives information about the geological processes that formed the rock.

Lineation in rocks is brought about by deformation, and can include such structures as slickensides, boudinage, rodded quartz and the intersection of two planar surfaces such as bedding and cleavage. It is typified by a parallel and linear orientation of mineral grains. This structural feature can indicate movement, producing slickensides, and rock extension caused by stretching, resulting in boudinage and rodding.

Foliation structures are planar, or roughly planar, surfaces formed in a rock as a result of deformation, often brought about by regional metamorphism. Foliation usually results from the alignment of minerals with a platy habit, such as micas, under intense compression. This is common in regionally metamorphosed rocks, and is shown by lenses and bands of different minerals.

Minerals can be rotated in some instances, to become orientated to a new direction, and new minerals can grow in the direction of foliation. A number of events can each produce foliation in a rock. Slaty cleavage, schistosity and gneissose banding are all types of foliation. These are new structures, found only in the deformed rock, not the original unmetamorphosed equivalent.

LOW-GRADE ROCKS

These rocks are formed at no great depth and on the margins of fold-mountain belts. Pressure is generally a more important influence than temperature, and this pressure produces structural features in the rocks created. Not many rocks are altered in conditions of low-grade regional metamorphism, usually only fine-grained sedimentary rocks and volcanic ash. The rock formed, slate, is very different from the rather structureless hornfels created from similar fine-grained rocks during contact metamorphism. Slate is the low-grade rock formed by alteration at low temperatures and pressures. Its main structural characteristic is cleavage (slaty cleavage), whereby the rock can be easily split into neat layers, a feature that has been used by the building industry for many years. The cleavage structure in the rock is produced by the parallel alignment of flaky minerals, including micas and chlorite, and new orientation of other particles in the rock. This slaty cleavage is referred to as foliation, and is a new arrangement of planar surfaces produced in a rock by deformation. Cleavage generally forms at right angles to the direction of maximum compression of the rocks. At this low grade of metamorphism, bedding and other structures may still be apparent, and the rock can contain distorted fossils.

▶ (top) Devonian slates from Valentia Island have been exploited for many years, the quarry first opening in 1816. Slate from here was used for roofing the Houses of Parliament, Westminster Abbey and many other large buildings. In the quarry, large blocks of slate can be seen, and the way they have broken displays the typical slaty cleavage structure of this low-grade rock.

Slate with cleavage, Ireland.

Cleavage surface of slate with pyrite, Cumbria.

▲ Slate frequently contains small porphyroblasts, discrete minerals embedded in the cleavage planes. Here, a cleavage surface is dotted with small (up to about 10mm/0.4in) crystals of pyrite, which have formed during metamorphism. The tiny black specks on the cleavage surface are also pyrite.

Bedding and slaty cleavage, Leicestershire.

▲ This slate, exposed at Charnwood Forest, was originally deposited as volcanic tuff, alternating with sandstone and siltstone, all of which are of late Pre-Cambrian age, dated at around 560 million years old.

The cleavage and bedding relationship is easily seen in this picture. The traces of the cleavage structures are shown by the thin vertical markings on the rock surface, and the bedding dips to the right. This cleavage affects all the rocks in the small section, both the finer and coarser layers. Radiometric dating of the micas on the cleavage planes shows that the structures in these rocks were produced by compressional forces at the end of the Silurian period, around 420 million years ago.

▶ (top) In contrast to the previous picture, here the darker, finer-grained slates show vertical cleavage, but the coarser sandstone beds lack cleavage. This is a typical feature of low-grade regional metamorphism, with more resistant rocks not being affected.

▶ (below) This abandoned quarry, in Borrowdale, Cumbria, shows the bedding/cleavage relationship in slates of Ordovician age. Originally fine-grained volcanic ash, these rocks were regionally metamorphosed during the Caledonian orogeny (period of mountain building) around 400 million years ago. The original bedding of the ash can be seen as distinct layers halfway up and above the small tree, the upper layer being the underside of a bedding plane. The slaty cleavage structures are the nearly vertical planes running through the rock.

Slaty cleavage, Leicestershire.

(below) Bedding and cleavage in slate, Cumbria.

Cleavage and bedding in slate exposure, Cumbria.

▲ Fine-grained greyish slate is exposed in many parts of Borrowdale. Here, near Grange, slate is seen on a hillside, and shows original bedding and cleavage, the latter produced by low-grade regional metamorphism and associated structural alteration. The bedding dips steeply towards the right of the picture, while the cleavage planes form near-vertical planar surfaces. These have been eroded into a slab-like pattern in the centre of the picture.

▶ (top) Very small-scale folding of the surfaces in slate can produce undulating features called crenulations. These are visible on the surface of the slate as tiny, wavy, ripple-like lineations. This type of microstructure is the result of a second phase of rock deformation, following the initial production of slaty cleavage. The crenulations form especially in fine-grained rocks such as slate and schist, which already have a well-defined cleavage or other foliation. These cleavage or schistosity surfaces are ideal laminations on which very small-scale crenulation folds can form. The small, discrete, 'rusty' porphyroblasts on the cleavage surface are weathered pyrite crystals.

▶ These highly inclined beds of Silurian greywacke and slate, exposed near Shap in Cumbria, show how beds with a different degree of competence are affected by low-grade regional metamorphism. The slate near the centre of the exposure, with the hand lens resting on it, has cleavage planes dipping at a high angle towards the left, while the more competent (resistant to alteration) beds on either side lack cleavage.

Crenulation on a cleavage surface, western Scotland.

(below) Cleavage, Cumbria.

Slaty cleavage, southern Scotland.

Slaty cleavage, South Wales.

▲ As has already been seen in slate from Charnwood Forest, slaty cleavage may not affect all the beds adjacent to one another. The more resistant layers (competent beds) often lack slaty cleavage, while the easily affected rocks are cleaved. Competent beds are able to withstand compressive forces more readily than weaker, incompetent beds. Cleavage of this type develops only when the original rock has suffered considerable compression, possibly as much as 30 per cent.

The degree to which a rock resists cleavage may depend on its mineralogy and structure. Two examples are shown here, one from southern Scotland, the other from South Wales. The cliff exposure from southern Scotland shows alternating cleaved and uncleaved layers. The coastal example from Wales shows highly cleaved beds resting on more resistant sandstones.

MEDIUM-GRADE ROCKS

In medium-grade regional metamorphism, a greater variety of rocks is altered than at low grade. This is because the temperature and pressure influences are greater, and especially because moderate temperatures are involved, whereas at low grade, temperature is hardly an influence. Most sedimentary rocks, including limestone and sandstone, as well as those of fine-grain size, and some igneous rocks can be altered by the conditions of medium-grade regional metamorphism. Also changed is the low-grade metamorphic rock, slate. The characteristic rock produced under these conditions is schist. New minerals, such as pyroxene, garnet and kyanite, may form.

The main structural feature of medium-grade rocks is a wavy foliation (schistosity), picked out by glittery layers of mica. The rock may break along these foliation surfaces, but does not break as easily as slate along the cleavage planes of that rock. Schist is a coarser-grained rock than slate, and in some cases original sedimentary structures such as bedding may still be visible. Medium-grade rocks are formed at greater depth in the Earth's crust and nearer the central regions of mountain belts than those of low grade.

▼ The wavy foliation structure typical of schist is clearly observed in this specimen. These foliation surfaces have a silvery sheen because of the presence of mica. The specimen is 20cm (7.9in) from side to side.

Schistosity, hand specimen.

Folded schist, Sutherland.

▲ Rocks formed during medium-grade regional metamorphism are frequently involved in folding. Here, in central Sutherland, Scotland, rocks of the Moine Schist have been folded in such a way that in the centre of the structure, compression has caused the folding to be far more extreme than on the margins.

▶ (top) This exposure near the Kyle of Tongue, on the north coast of Scotland, shows medium-grade schist derived from sandstones. Much mica has developed during metamorphism, giving the rocks a silvery sheen, but some of the original sedimentary structure remains, and farther along the exposure cross-bedding can be recognised.

▶ A detailed picture of schist shows the dominance of mica, which follows the wavy foliation structures. This schistose foliation is emphasised by the folded quartz vein. The brownish porphyroblasts are garnet crystals, which are characteristic of the medium grade of regional metamorphism. The whole field of view is about 10cm (3.9in).

Schist, Sutherland.

Folding in schist, Sutherland.

▶ (top) This picture shows intensely folded gneiss from the Lewisian basement of Assynt, Sutherland, Scotland. The scale is about 60cm (24in) across the field of view. The characteristic banded structure of this high-grade metamorphic rock is well displayed, with quartz- and feldspar-rich material forming the paler layers, and hornblende and biotite layers making up the darker bands. This is an orthogneiss, formed by the high-grade alteration of igneous rock. The intense, very tight folding suggests that the rock may well have been in a relatively plastic state during metamorphism.

HIGH-GRADE ROCKS

High-grade regional metamorphism occurs at very high temperatures and pressures, conditions found deep within the roots of mountain chains, especially during orogenic activity brought about by plate tectonic movement. As well as the temperature and pressure influences, circulating fluids are important in introducing and removing mineral material from the rocks. Virtually all pre-existing rocks, even igneous rocks such as granite and dolerite, are altered at this level of metamorphic activity. Indeed, much of the high-grade gneiss that covers extensive parts of the continental shield areas is granitic in origin. Gneisses that were previously igneous rocks are termed orthogneisses. Those believed to have originally been sedimentary are paragneisses.

The main structural feature of the high-grade metamorphic rocks is a definite banding, in which different mineral groups are concentrated. One of the typical features of gneiss is the alternating dark and light bands. The darker bands have a predominance of amphiboles, pyroxenes and dark mica, while the paler bands are made of quartz and feldspar. When there are narrow discontinuous bands and layers of pale acidic and darker basic material running through a mass of gneiss, the term lit-par-lit gneiss is applied. The name migmatite is used for gneisses that have a mixing of acid and basic material. The segregation of minerals is thought to be a response to the extremes of temperature and pressure.

Folding of banded gneiss, Sutherland.

Gneiss landscape, Sutherland.

▲ Here, an exposure of basement gneiss on the west coast of Sutherland, Scotland, shows the typical hummocky landscape produced by this most durable rock. There is an indication that the banded structure of the gneiss has influenced the weathering of the rock surfaces, as the paler, quartz-rich, more acidic bands are prominent on the exposed rock surfaces.

Gneiss exposure, North Uist.

▲ Gneiss is a dominant rock type on many of the Outer Hebridean islands; indeed, the term 'Lewisian gneiss' comes from its predominance on the Isle of Lewis. Here on the west coast of North Uist, high-grade gneiss is exposed, showing its typical banded structure. The characteristic dark and pale banding is seen in the small vertical slope in the centre, while to the left there are more discrete, dark, basic or ultrabasic masses of rock, which may result from the break-up of a basic dyke. There is also evidence of the gneiss being tightly folded. To the extreme right of the exposure, a pink-coloured band of acid, feldspar-rich pegmatite is exposed. This intrusion has a width of around 30cm (12in). Further masses of pinkish pegmatite occur on top of the gneiss exposure. The complexity of the geology of these basement rocks is illustrated by the gneiss and intrusion of both acid and basic rocks.

▶ (top) This exposure of basement gneiss shows in the lower parts typical banded rocks, dark basic material alternating with pinkish acid layers. Through the centre of the picture, running across the exposure, is a dark greenish-black mass of rock, which is composed mainly of amphiboles, particularly hornblende and actinolite. Though it is now rather irregular in outline, and somewhat fragmented, this dark rock may well have been a dyke intruded into the basement rocks prior to later metamorphic events. A rounded 'ball' of actinolite can be seen just above the centre of the picture. The scale is given by the small plants of thrift, flowering to the left of centre.

Basement gneiss, Sutherland.

▼ There is a strong contrast in this small exposure, about 2m (6.6ft) across, between the typical banded structure of gneiss and many discrete, dark, ultrabasic rock masses. These are generally rounded in outline, and the banding of the gneiss can be seen to 'flow' around them. The fragmented, dark-coloured rock is referred to as agmatite. It seems that when there is a mixing of acidic gneiss and basic or ultrabasic rock, the dark rock is pervaded by the acidic gneiss and broken into these isolated, rounded masses. Eventually the ultrabasic material may be incorporated into the gneiss and lost in its banded structure.

Agmatite, north-west Scotland.

Stratification in the canyon of the San Juan River, Utah.

▲ Stratification is the most characteristic structure of sedimentary rocks. This feature is often best seen where erosion has cut through the strata. Here, in the Goosenecks State Park, Utah, USA, the San Juan River has eroded deep meanders through various horizontally bedded rocks. The erosion is probably due to gradual uplift of the plateau surface. Such deeply cut meanders are locally known as 'goosenecks'. At the base of the canyon, forming steep cliffs, there are fossiliferous marine limestones deposited during the Carboniferous period. Above these rocks are younger Carboniferous strata, including shale, limestone and sandstone, which form gentler slopes covered with loose scree.

SEDIMENTARY ROCKS AND THEIR STRUCTURES

Sedimentary rocks are formed on the Earth's surface in environments, many of which can be readily studied, unlike those in which the majority of igneous and metamorphic rocks are created. Sedimentary environments range from river systems and their deltas to the continental shelf and abyssal ocean plains. Sediment is also formed on the dry land surface, especially in the vast sandy deserts. The interpretation of many ancient sediments can therefore be carried out with a good degree of certainty. Essentially they are composed of particles derived from earlier-formed rocks or chemical and biological material. One of the fundamental principles used in their interpretation is that of uniformitarianism, first proposed by James Hutton (1726–1797), a pioneering Scottish geologist. His idea that 'the present is the key to the past' suggests that rock-forming processes we can see and understand today are the same as those that occurred in the past.

An important structure of sedimentary rocks is that of stratification, also known as bedding. Every stratum (layer or bed) of sediment is deposited upon a previously formed stratum. Bedding planes can easily be recognised in an exposure, dividing one layer from another, and distinguish sedimentary rocks from igneous and metamorphic rocks. Sedimentary rocks do not always exist in neat horizontal layers, but often have fold structures, or are tilted, sometimes at high angles from the horizontal. These structures have developed after formation of the rock. It is generally thought that the sediment from which these rocks are created was deposited in horizontal layers, except in certain situations, for example, where scree deposits occur on slopes, often at angles of up to 40°. Also, these strata can be laterally continuous over a wide area, often with a consistent thickness.

The formation of one sedimentary bed upon another is the basis of the science of stratigraphy. This is the interpretation of the sequence of rocks, and the reconstruction of past environments. The geological 'law' of superposition says that younger beds lie on older ones. This holds true unless overturning by folding has occurred. In order to determine if the strata being studied are overturned or otherwise altered in their attitude, way-up criteria can be applied. Here, sedimentary structures are important. Also, a bed containing fragments of another rock must be younger than the rock from which the fragments were derived.

Sedimentary rocks differ from most igneous and metamorphic rocks in that they may contain a range of fossils. These are evidence of previous life and its evolution on Earth, and are a great help to stratigraphers. Based on the principles first pioneered by the English canal engineer William Smith (1769–1839), it is possible to recognise certain sedimentary beds by their included fossils and place them in a relative time sequence. Fossils known as zone or index fossils, such as the Jurassic ammonites, are ideal for this work, as they had a wide geographic range and evolved rapidly into different species, each one only occurring for a relatively short time. Thus a very neat stratigraphic sequence has been established using these fossils, and strata can be correlated over wide areas.

Three main types of sedimentary rocks are recognised: detrital, organic and chemical.

▶ (top) In Arizona, USA, the Colorado River has eroded through 1,800m (5,900ft) of strata. These virtually horizontal sedimentary beds have gradually been uplifted as part of the Colorado Plateau over the last 65 million years. The rocks at the base of the Canyon are Pre-Cambrian in age (about 1,500 million years old), while those on the upper levels are 230 million years old, formed during the Triassic period.

▶ Sedimentary bedding is well illustrated here, at Hunstanton in Norfolk, by red and overlying white chalk, both of Cretaceous age. The distinct surface that separates the two types of chalk can be followed along the cliff line. White chalk lacks detrital material and is almost pure calcite (calcium carbonate), while red chalk has a high percentage of iron oxide, providing the red colour. The cliffs here are about 18m (59ft) high.

Stratification, Grand Canyon, Arizona. Below: Stratification, Norfolk.

Steeply dipping strata, Menorca.

▲ Sedimentary beds, originally formed on horizontal surfaces, are often involved in earth movements. Here, the alternating brown sandstones and grey shales of Carboniferous age are clearly visible, but they have been tilted at a very high angle. The cliff exposure, near Es Mercadal, Menorca, is about 4m (13ft) high.

▶ (top) The base of this 3m (9.8ft) high cliff, at Widemouth, Cornwall, shows very obvious bedding planes, dipping towards the right of the view. The dip of the strata may appear to be different when viewed from other positions. Sandstone beds, with thin shales between them, lie below a massive sandstone with few bedding planes visible. This rust-coloured sandstone is rich in iron oxides, which have formed rounded structures called Liesegang rings.

▶ These iron-stained, concentric, ring-shaped structures are thought to be of diagenetic origin and form in porous sedimentary rocks by the rhythmic precipitation and dissemination of iron oxides from fluids in saturated rock. The field of view is about 1m (3.3ft) across.

Dipping strata, Cornwall.

Liesegang rings, Cornwall.

DETRITAL SEDIMENTARY ROCKS

The detrital rocks are formed from particles worn, by weathering and erosion, from pre-existing rocks. These particles are transported by water and the wind, to be deposited in beds, often in the sea, when the transporting agent loses its force. These rocks are classified according to the size of their constituent grains.

The coarse-grained sediments, with a particle size over 2mm (0.08in) in diameter, include breccias and conglomerates. The difference between these two rocks is that conglomerates are made up of rounded, often water-worn fragments, while breccias contain angular particles, which may result from scree deposits on steep hillsides. These rocks can have a variable mineral content and they usually contain many rock fragments.

The medium-grained rocks, with particles between 0.06 and 2mm (0.002 and 0.08in), are essentially the sandstones. These rocks contain a very high percentage of quartz, but other minerals, such as feldspar and mica, may be present. As the percentage of quartz increases, so the rock is said to be more mature. Quartz is a very durable mineral and will stand erosion, weathering and the rigours of transportation better than most other minerals. Arkose, a type of sandstone containing feldspar, is less mature than a pure quartz sandstone, and has possibly been deposited quite rapidly in an environment where feldspar is not readily altered to clay.

The fine-grained (less than 0.06mm/0.002in) detrital sediments are shale and clay. Mudstone is similar to shale, but less well bedded or laminated.

Structures in the detrital rocks include markings on the bedding planes and structures within the beds themselves, such as cross-bedding, desiccation marks, ripple marks, groove casts and washouts.

ORGANIC SEDIMENTARY ROCKS

Organic sedimentary rocks include many types of limestone, especially those formed of broken fossil material such as bivalve and brachiopod shells. Coral can be a constituent of such rocks, and may be in large reef structures. These limestones are often named according to their organic content, for example, shelly limestone or coral limestone. Coal, being formed of plant material, is an organic sedimentary rock. These sediments contain many structural features related to their origin, including algal mounds and coral colonies.

CHEMICAL SEDIMENTARY ROCKS

Chemical sedimentary rocks are those for which the precipitation of chemicals and minerals is an important rock-forming process. Some limestones are chemically formed. Oolitic limestone is produced by the precipitation of calcite around small nuclei such as sand grains or shell fragments. The rounded ooliths measure only about 2mm (0.08in) or less in diameter, and are deposited in strata as oolitic limestone, which sometimes shows cross-bedding. Their formation is aided by constant agitation of warm seawater. Some ironstones are deposited chemically, as are the group known as evaporite rocks. In certain situations, such as drying marine lagoons and inland drainage basins, minerals are precipitated in beds as the water evaporates. Rock salt, gypsum rock and potash salts, all of great economic significance, form in this way. They are commonly interbedded with marl, a calcite-rich clay.

FROM SEDIMENT TO ROCK

Diagenesis is the term used to encompass the various processes and changes that take place enabling loose sediment to be converted into a more durable rock. When sediment has been deposited it is often covered by many younger layers. Considerable weight of sediment causes compaction, which tends to affect the finer-grained mudstones and clays to a greater degree than coarser rocks. This process helps to remove pore spaces between the constituent grains in the sediment.

Where the grains rub against each other, they are moved closer together and pressure solution can occur. Fluids, often water, held in pore spaces can be expelled. Buried sediments may contain seawater, which is chemically active and can move minerals in solution. These minerals, including quartz, calcite, haematite and limonite, can be redeposited around the grains in the sediment, acting as a cement. Nodules and concretions found in sedimentary rocks are structures closely related to diagenetic processes.

JOINTING STRUCTURES IN SEDIMENTARY ROCKS

Fractures in rocks, which, unlike faults, show no vertical displacement on either side of the fracture, are called joints. They are very common features in many types of sedimentary rock, especially those of medium and coarse grain size. Joints can occur in parallel sets, or several sets may intersect, giving the rock a stepped and block-like structure. A series of joints that are parallel to each other is referred to as a joint set. The term joint system is applied to a number of joints that cut through one another.

Joints in sedimentary rocks can be the result of drying out, causing a certain degree of shrinkage. While at depth, a rock layer may be under considerable pressure, which is reduced by unloading. After the weight of overlying material has been removed by erosion, expansion of the rock can lead to the formation of joints.

Joints can have an important bearing on the subsequent weathering and erosion of the rock and the landforms that may develop. The presence and circulation of fluids within the rocks of the crust can be greatly influenced by joints. The movement and accumulation of natural gas, and hydrothermal fluids containing ore-forming chemicals, may be controlled by joint systems in sedimentary rocks. Water flowing on the surface over jointed rocks may be channelled underground. At the coast, joint systems have a considerable impact on landform development, especially the formation of cliffs, arches, stacks and caves.

▶ (top) Carboniferous limestones, with beds dipping landwards, form this classic coastal arch and associated stack on the coast of Wales. Landforms such as these are often associated with structures in the sedimentary rocks of which they are made, and here joints are influential. In the limestone to the right of the arch, two important joints are clearly seen. These are being actively eroded. Initially the coastline is one with caves developed along joint systems. The caves are gradually enlarged to form arches, such as the Green Bridge, and subsequent collapse produces stacks.

▶ Dipping Torridonian sandstones are seen here to have vertical joints, which are especially noticeable to the left of the caves, in the upper part of the cliff. There are further joints to the right of the caves. The cave

Green Bridge of Wales, Pembrokeshire.

Caves and joints, Sutherland.

development has been influenced by these joints, which are weaknesses in the strata, readily exploited by marine erosion and by water running down the cliff from above.

▼ Tors are not only features of granite. Here, at Brimham Rocks in North Yorkshire, tors have developed in gritstone (coarse, often pebbly, sandstone) as a result of its bedding plane and vertical joint structures. Certain bedding planes have been weathered more deeply than others because of their different types of cementation. These have been widened by weathering, and it is thought that extensive freeze-thaw action towards the end of the last glaciation was responsible for much of this joint and bedding plane enlargement. The gritstones show cross-bedding, related to the alluvial environment in which they were deposited.

Gritstone tor, North Yorkshire.

Joint system, Tyne and Wear.

▲ This Carboniferous sandstone wave-cut platform, at Whitley Bay, shows two sets of intersecting joints forming a joint system. These joints allow the rock to be eroded, forming large rectangular blocks. On the seaward margin of the wave-cut platform these blocks are becoming detached as the coastline is actively eroded.

JOINTS IN LIMESTONE

Many theories have been put forward for the origin of the extensive joint systems in limestones. As well as being produced by shrinkage of the sediment, their origin may be related to the unloading of overlying sediment and subsequent expansion of the limestone mass. In many limestones, joint systems are enlarged by chemical weathering. The rock is mainly composed of calcite (calcium carbonate), which, when in contact with groundwater and rainwater (both of which are slightly acidic), is altered to soluble calcium bicarbonate and so carried away in solution. Thus the joints in the rock are enlarged and the rock becomes permeable. The development of aquifers and underground water systems can be a direct result of the rock's permeability and the chemical weathering of limestone joint structures. This has considerable influence on the rock and its associated landforms.

Karst landscape, western Ireland.

▲ A characteristic feature of many limestone areas is the lack of surface water and the wide expanses of exposed bare rock. Here, in County Clare, Ireland, the joint structures in Carboniferous limestone give the rock its permeability. Because of the lack of water, soil development and plant growth is very restricted. The bare rock landscape, often containing sinkholes and cave systems, is referred to as 'karst' scenery. The term is taken from a German name for a limestone area with these features near Trieste, Italy.

▶ Limestone pavements develop on jointed limestones in many areas. Some form after the removal of overlying material by ice sheets, but pavements also develop beneath a covering of regolith. After retreat of the ice sheet, the bare limestone has high permeability because of its enlarged joints, and surface water and soil formation are at a minimum. As the joints are enlarged, a system of blocks of limestone is created. The vertical gaps (joints) are referred to as 'grikes' and the blocks as 'clints'. The limestone pavement above Malham Cove in Yorkshire shows these features, and the slightly rounded edges of the weathered limestone developed in response to chemical weathering.

Limestone pavement, Yorkshire.

Sinkhole, North Yorkshire.

▲ When joints are considerably enlarged, collapse of the limestone may occur. Large vertical fissures can then develop, and water with mechanical erosive power can flow underground. Sinkholes formed in this way are a typical feature of karst landscape. Underground drainage systems often lead off from limestone sinkholes.

▼ The joints in many limestones, here in Cambrian limestones in northern Scotland, allow water to permeate the rock. When impermeable strata occur below the limestone, a line of springs can arise, where underground water flows out on to the surface, especially at times of high rainfall.

Springline, Sutherland.

STRUCTURES RELATED TO SEDIMENTARY DEPOSITION

Stratification, or bedding, is probably the most apparent depositional structure of sedimentary rocks. Depending on the environment and the movement of water or air when the sediment is being deposited, other structures may be created. Often there are angled surfaces within a bed (cross-bedding), formed by a moving current during deposition. The surfaces of sedimentary layers may have ripple marks, as seen on a modern beach, and when sediment settles it may be graded, with the coarser particles falling to the depositional surface first.

CROSS-BEDDING

Cross-bedding is a very common sedimentary structure, especially in sandstones and some limestones. It is strongly related to flowing currents of water, or wind, and the deposition of sediment in such a moving system. In the past, the terms current and false bedding have been used for cross-bedding, but these are no longer acceptable.

This structure is quite easy to recognise in an exposure, but may be more difficult to interpret. Between the base and top of a specific bed, cross-bedding appears as a series of inclined surfaces, parallel to each other. Each cross-bedded unit is called a set and several sets make up a coset. The inclined layers have been deposited by a flowing current, and dip down in the direction in which the current flowed. In a cross-bedded rock, the sloping layers are called foresets. These are generally composed of sand-sized particles and make up the major part of the deposit.

Cross-bedding is created by the movement of sediment within structures, including ripples, sandbanks, dunes and deltas. As the sediment is moved by a current, erosion occurs on the stoss (upstream) side and deposition takes place on the lee (downstream) side. In this way, the feature moves in the direction of the current and its internal structure is one of cross-bedding.

Cross-bedded strata can take on various shapes. The angle at which they are exposed is critical in their understanding and interpretation. When the cross-bedded sets have roughly parallel top and bottom surfaces and the layers of the foreset beds dip at the same angle, the structure is called tabular cross-bedding.

This is formed by sand waves with straight tops. In trough cross-bedding, the sets have a basin-shaped structure, with top and bottom surfaces that are not parallel. This type of cross-bedding is produced by dunes with curved crests.

Tabular cross-bedding, Northumberland.

▲ These sandstones of Carboniferous age show cross-bedded structure, with their foreset beds dipping to the left. The top and base of each unit of cross-bedding are roughly parallel. The beds were deposited in moving water currents, possibly associated with a large delta. The currents flowed from right to left, and the sediment spilled over the previously deposited layers. Weathering has picked out certain sandstone beds, accentuating the overall structure.

▶ (top) A Cambrian quartzite exposure in north-west Scotland shows herringbone cross-bedding. Within a short vertical distance, the dip direction of the cross-bedding changes. At the top of the picture, the beds dip to the right, in the centre there are beds dipping left at a low angle, and the lowest beds dip to the right. Some beds are picked out in brownish colouring possibly caused by iron staining. These beds are of shallow marine origin. The rapidly altering dip direction of the cross-bedding suggests changing water currents and possible tidal influence.

Herringbone cross-bedding, Sutherland.

▼ This exposure is found high on the slopes of a mountain called Quinag, in north-west Scotland. It shows pale grey Cambrian quartzite with cross-bedding structures. The foresets dip to the right, but the cross-bedding in the centre on the cliff shows a change of dip at its base. Here the beds curve and their dip lessens, as they meet the underlying surface. This type of cross-bedding is useful in stratigraphy as a 'way-up' criterion, as the concave surfaces are uppermost when formed.

Cross-bedding, Sutherland.

Cross-bedding, Orkney.

▲ On the coast of Orkney, at Yesnaby, sandstones of Devonian age are exposed. These strata show extensive cross-bedding, with different beds dipping in different directions; an area of herringbone cross-bedding can be seen towards the base of the main cliff. This suggests changing directions of current flow during deposition, which was possibly by braided river systems. The stack has a small arch eroded at its base, and is in a most precarious state.

▶ (top) These Carboniferous sandstones show repeated cross-bedded layers in a cliff exposure about 1m (3.3ft) high. The sloping foreset beds dip to the right between the top and bottom surfaces. This direction of dip indicates that the water in which the beds were deposited was flowing towards the right, probably in river systems, meandering and depositing sediment in a deltaic environment.

▶ The moving currents that produce cross-bedded sediments frequently make erosive structures such as troughs in previously deposited sediment. Troughs have curved lower surfaces. Here, a trough cuts through cross-bedded sandstone. Because this section cuts the trough vertically at right angles to its length, cross-bedding of sediment in the trough is not apparent. The exposure is about 1m (3.3ft) high.

Cross-bedding, Northumberland.

Trough cross-bedding, Tyne and Wear.

Dune bedding, County Durham.

▲ Dune bedding, produced by the large-scale movement of sand under the influence of the wind, is usually on a bigger scale than water-deposited cross-bedding. The cross-bedded units are often wedge-shaped, and individual units are measured in metres, as opposed to the centimetres of water-formed units. The foresets are a result of sand spilling over the lee side of dune structures. These two pictures show dune bedding in sandstones of Permian age in County Durham, UK. Magnesian limestone is visible above the dune-bedded sand in the wider view.

RIPPLE MARKS

Ripple marks are a well-known feature of many water- and wind-influenced environments in which sand is being deposited. They are common along the coastline and around lakes and rivers. Sandy desert regions have ripple marks on their surfaces, and dune systems are, in many respects, mega-ripples, with a similar mode of origin, though they are on a much larger scale. The actual shape and size of a ripple mark depends on the type of sediment being moved and deposited (usually fine sand or silt) and the nature and speed of the depositing agent.

Ripple marks may have straight or sinuous crests, with a height of about 3cm (1.2in). The ripples are usually spaced up to about 30cm (12in) apart, and when being actively deposited, ripples migrate down current and may climb over one another. If ripple marks are formed by consistently moving currents, they have a gentle stoss side and a steep lee side, giving them an asymmetrical profile, with sediment moving up the stoss side and being deposited on the down-current lee side. Ripples resulting from an oscillating current are more symmetrical in cross-section.

Ripple marks, Sutherland.

▲ Ripple marks are seen on sandstone bedding planes of Pre-Cambrian age, belonging to the Torridonian supergroup. These structures suggest shallow water currents, possibly those of migrating river systems.

The sediment deposited in ripple-mark formation has an internal structure of cross-lamination, a type of small-scale cross-bedding, related to the movement of the sediment of which the ripple marks are formed. Ripple marks are common in sandstones and other fine- to medium-grained sedimentary rocks, and are often encountered on bedding planes. They can be of use as way-up criteria.

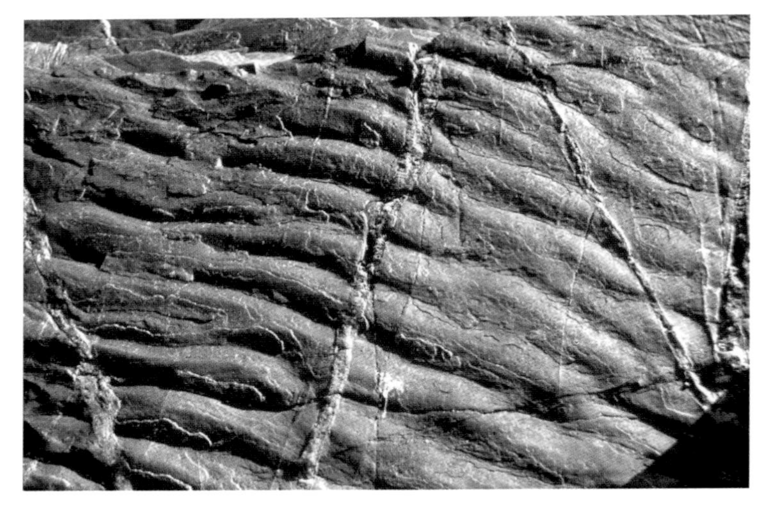

Ripple marks, Cornwall.

▲ This bedding plane in rock of Carboniferous age, exposed at Bude in Cornwall, shows ripple marks that are more or less symmetrical with rounded crests. In contrast to asymmetrical ripples, these are formed by an oscillating current, whereby the water moves backwards and forwards over the sedimentary surface.

▶ (top) It is believed that sedimentary structures such as ripple marks were formed in the same way in the distant past as they are at the present day. These beach ripples have parallel crests and are sinuous along their length. It can be appreciated here how the sandy sediment has been moved by tidal currents, in the same way that larger structures can migrate in advance of a current of wind or water. In this shallow marine environment, each tide alters the pattern of the sediment ripples.

Modern beach ripples, Sutherland.

Sand ripples, Gobi Desert.

▲ This picture, taken in the Gobi Desert, shows parallel, sinuous ripples on the desert surface. The ripple crests are somewhat complex, but suggest that the dominant wind direction is from the left, as the more gentle stoss sides of these asymmetrical ripples dip in this direction. The dune system in the background indicates the same wind flow, with steep lee slopes to the right. Some of the dunes here have a slightly crescentic pattern, with their margins extended.

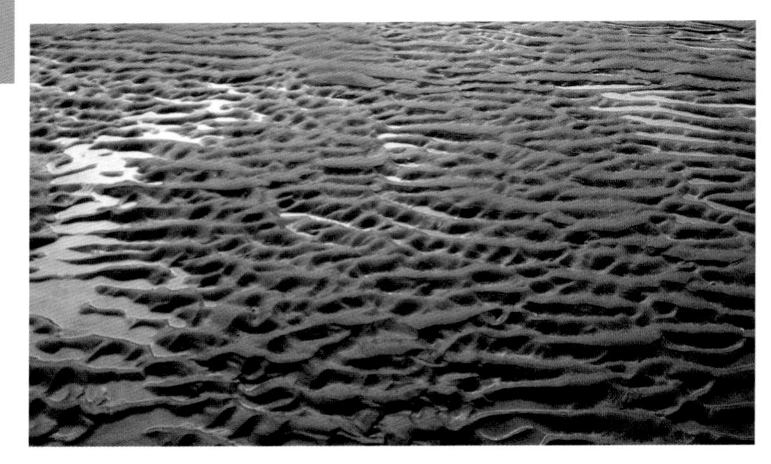

Interference ripples, North Yorkshire.

▲ On this modern sandy beach, two sets of ripple marks have been formed. The crests of the dominant set run almost from left to right. The gentle stoss sides dip towards the top of the picture and the steeper lee sides to the bottom. A second set of ripples, seen in the troughs between the main ripples, has crests running almost at right angles to the principal set. Two water currents have been involved in the formation of this interference pattern, the first flowing towards the camera and a second one from left to right.

▶ (top) This type of very small-scale cross-bedding is known as cross-lamination. Here, in rocks of Carboniferous age, a number of cross-laminated sets can be recognised. The structure is produced by wave action creating sediment ripples, which rise above each other as they move downstream. The crests of the ripples are pale and the troughs are infilled with darker sediment. The term climbing ripples can be applied to this structure.

▶ The deposition and movement of sediment on a small scale can produce many sedimentary structures including ripple drift bedding. Here, in strata of Carboniferous age, reddish-coloured muddy sediment occurs in the ripple troughs. This is frequently cross-bedded, and the concave upwards shape is a useful indication of way-up.

Climbing ripples, Tyne and Wear.

Ripple drift bedding, Tyne and Wear.

GRADED BEDDING

Sedimentary deposition can produce strata that are referred to as graded beds. When such a stratum is studied in detail, it can be seen that the sediment at its base is coarser than that at the top, and there is a gradual 'fining-upwards' in the stratum. This grading may be, for example, from sand to shale or clay. The internal form of a graded bed often lacks structure. It is commonly the case that graded beds occur in cycles, with many individual beds deposited one above another. The base of each graded bed may make a sharp contact with the top of the one below, and the coarser particles at the base of a graded bed can cause indentations and marks in the fine, often shaly, top of the underlying bed.

Grading of beds can occur in fluvial situations, where river currents slow down and coarse sediment is deposited first, with gradually finer sediment settling above it. In lakes, there may even be a seasonal deposition of sediment, with finer material in the drier season and coarser material brought in when precipitation is higher.

A classic situation in which graded beds are formed, however, is where turbidity current deposition occurs. Turbidity currents are made denser than the water around them by their sediment content. This sediment can be a mixture of coarse and fine material, with a considerable range in size. Such a current may be triggered by a minor earthquake or slumping on the margin of the continental shelf, and the slurry of sediment disturbed in this way can flow down the continental slope on to the ocean floor. It is commonly thought that submarine canyons (deep, steep-sided valleys cutting though the continental slope) may have been partially eroded by turbidity currents. Off the east coast of the USA are canyons over 3,000m (9,840ft) deep, and many others on a similar scale have been mapped around the world.

A submarine canyon also helps to shape a turbidity current as it flows into deeper water. It has been calculated that such currents can flow at 100kph (62mph) or more. These currents are a transport system by which coarse sediment is taken into deep ocean water, and they also provide a mechanism for the formation of graded beds. The coarse sediments are deposited first and as the flow slackens in velocity on the less steeply sloping sea bed, gradually the finer silt and mud is deposited, to create a graded bed. As these currents tend to flow in cycles, many graded beds are deposited one above another, each representing the sediment laid down by a single current. The graded beds formed in this way are referred to as turbidites.

Graded beds are of great value to stratigraphers. Because of the deposition of such beds, with coarse sediment at the base and fine at the top, and their cyclic or repetitive nature, they can be relatively easy to recognise in a given sequence of rocks. It is generally possible to work out the way-up of graded beds and hence the stratigraphic sequence.

Turbidites, Devon.

▲ These tilted Carboniferous beds, at Hartland Quay on the coast of North Devon, show alternating sandstones and mudstones. The sandstones make up the thicker, more prominent parts of the section, as they have resisted erosion; the mudstones are thinner and often recessed. One relatively thick mudstone has been eroded to form the 'tunnel' through the exposure. The repetitive nature of these beds, as well as their lithology, suggests that they are turbidites, formed by an ocean floor turbidity current. Each current deposits sand first, followed by finer-grained mud.

Turbidites, Cornwall.

▲ The repeated coarse- and fine-grained beds of these typical turbidites can be seen on this 5m (16ft) cliff face. Different turbidity currents do not necessarily deposit the same amount of sediment, and the idealised upward fining sequence is not always present.

Inverted turbidites, Cornwall.

◀ Turbidite sediments, formed from deposition of the slurry of sediment carried into deep marine water by turbidity currents, usually show graded bedding. This structure can be invaluable when trying to determine the way-up of the beds. Here, on the north coast of Cornwall, repeated graded beds of Carboniferous age, about 50cm (20in) thick, are exposed. It can be seen from the grading that these beds have been upturned. The bed to the left of centre shows pale grey sandstone on its right-hand side. This takes up about half of the bed and grades to the left into darker shale and mudstone. Initially, when these beds were formed, the younger ones on the left were above those on the right.

STRUCTURES RELATED TO THE EROSION OF DEPOSITED SEDIMENT

Sedimentary rocks are formed in a great many environments, though most of the sediment that becomes part of the geological record is probably of marine origin. When this sediment is deposited, it is initially wet and unconsolidated, and can be readily eroded by currents and objects moved over its surface. Marks of many types may occur in this sediment surface, the shape and size of which depends, to a great extent, on the object causing the erosion, its speed, and the coherence of the sediment.

▼ Flute casts can form individually, or in groups, as here in Ribblesdale, Yorkshire. They are commonly found as positive structures on the underside of a sedimentary bed. Originally, the flute formed as a depression in a relatively soft and unconsolidated muddy sediment surface. Irregularities in the surface caused the flow of sea-bed water currents to become locally erosive, and a streamlined hollow then developed. The rounded, rather steep nose of the flute cast faces upstream. Preservation of the flute results from its infilling by coarser sediment. Erosion of the sedimentary rock, often a considerable time after the original deposition, tends to remove the shale or mudstone in which the flute was formed. The more competent sandstone bed, on the underside of which the flute cast exists, remains.

Such structures are commonly associated with turbidity current deposition, and are very good indicators of the direction of current flow. The length of individual flutes can vary from 5cm to 50cm (2in to 20in), and they may be between 1cm and 20cm (0.4in and 7.9in) wide. The field of view here is about 25cm (9.8in) across, and this picture is of the underside of a bed, the softer sediment in which the original flute was made having been eroded.

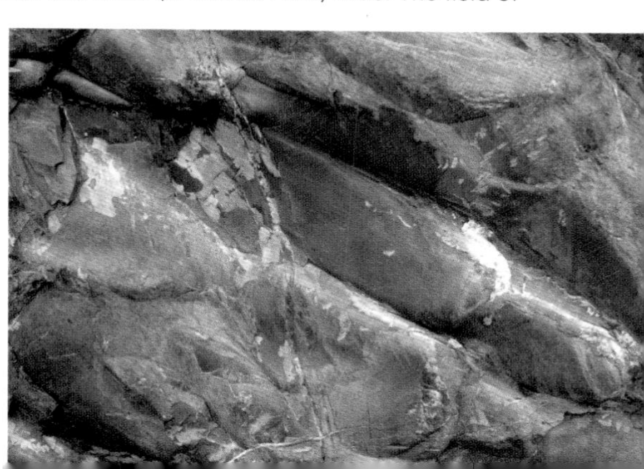

Flute casts, Yorkshire.

Grooves and other marks are formed initially as hollows, but when new sediment is deposited on the eroded surface, these hollows are infilled and so preserved. Eventually, when the sediment becomes rock, erosional structures can be found as positive features, or casts, on the bottom, or sole, of the bed that filled in the original hollows. It is often the case that the eroded and marked sediment is fine-grained, and the later infilling sediment is coarser. Sole structures are commonly associated with sedimentation that has occurred in episodes involving deposition and erosion. These structures are frequently found in turbidite beds, and are valuable as way-up criteria. Certain erosional structures, such as washouts, can be observed in vertical exposures.

▶ (top) This example of flute casts is from Shap, Cumbria, in shale associated with greywackes of Silurian age. The rounded nose of the flutes is to the left, suggesting that the current that formed them flowed from left to right. Above them are some small bounce marks.

▶ Small objects, such as sediment grains, pebbles and even fish vertebrae, may be bounced by currents over the sea-bed sediment surface, making indentations, especially if the sediment is fine-grained and muddy. The prod and bounce marks resulting from this activity are referred to as tool marks. The fragment (tool) making a prod mark scrapes into the sediment at a relatively high angle, and is then pulled out nearly vertically by the water current. This produces a rather blunt end on the downstream side. Prod marks are discontinuous casts found on the underside of bedding planes. They have an asymmetrical shape, often with a triangular outline. The example here is from Carboniferous rocks at Widemouth, in North Cornwall, and shows a series of prod marks, just above the coin. Also on this bedding plane are longitudinal groove casts.

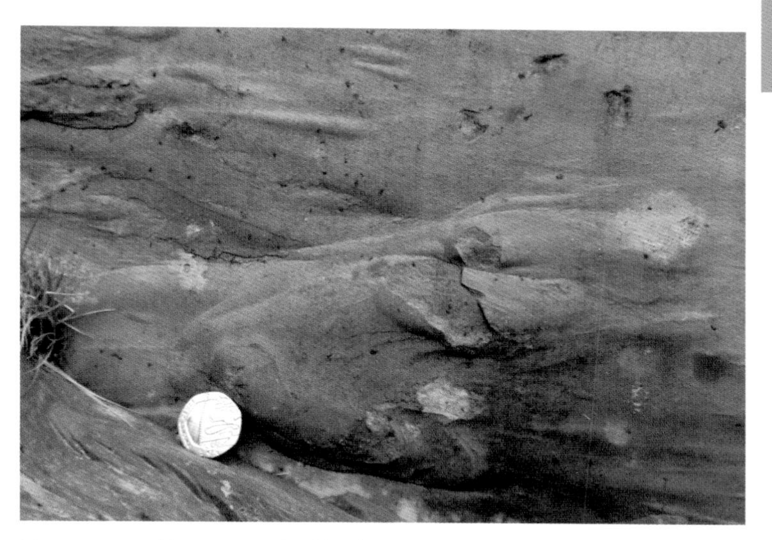

Flute casts and bounce marks, Cumbria.

Prod marks and groove casts, Cornwall.

Bounce marks, Cumbria.

◀ Bounce marks contrast with prod marks in both their shape and mode of formation. The object making the bounce mark hits the unconsolidated sea-bed sediment at a lower angle than in the production of a prod mark and so makes a less sharp impact, before rising away at a low angle. The resulting structure is therefore more symmetrical. Bounce marks are usually preserved as ovoid and elongate casts, with similarly shaped ends. It is rarely possible to identify the actual tool that created the bounce mark, but small rock fragments are generally held to be responsible. A number of bounce marks may occur close together, as on this shale bedding plane of Silurian age from Shap, in Cumbria. This example, in common with many tool marks, shows the positive infillings (casts) of bounce marks on the underneath (sole) of a steeply inclined bedding surface.

▶ Another type of tool mark often associated with turbidite beds is produced when an object is dragged along the surface of sediment that is sufficiently cohesive to be marked. As with other tool structures, the preservation of the groove depends on it being infilled by coarser material. Essentially, a groove cast is a longitudinal ridge found on the underside of a bed, commonly of sandstone. The two examples shown, both from North Devon, are on the base of turbidite sandstone beds. The lower one clearly shows groove casts striking in a number of directions, suggesting a variation in the direction of current flow. The cross-cutting relationships indicate the sequence of formation.

Groove casts, Devon.

Groove casts, Devon.

▲ All types of sole structures are of value to stratigraphers, as they indicate the way-up of the beds. This allows the relative age of the strata to be determined. The groove casts shown in the previous picture are exposed high up on the underside of a bedding surface to the left of the cave.

▶ (top) These sole marks were created by the scouring action of a turbidity current on a relatively soft sediment surface. The ridges, which are usually more or less parallel, are separated by sharp furrows. This structure can give evidence of the current direction. The width of this exposure is about 15cm (5.9in).

▶ Channels occur in a number of sedimentary environments, but those that form in deltaic situations are possibly the best preserved. A stream meandering across a delta top can cut a washout channel through soft, previously deposited sediment, and such channels may be preserved by later sedimentary deposition. Here, rocks of Carboniferous age are exposed on a small cliff about 5m (16ft) high. This concave structure is a cross-section of a small stream channel that has been eroded into cross-bedded sandstones. The base and sides of the channel are sharp and erosive. Massive (unbedded) sandstone fills the channel. The lack of structure in the channel sandstone may be a result of the collapse of the banks and a slurry of sand falling in.

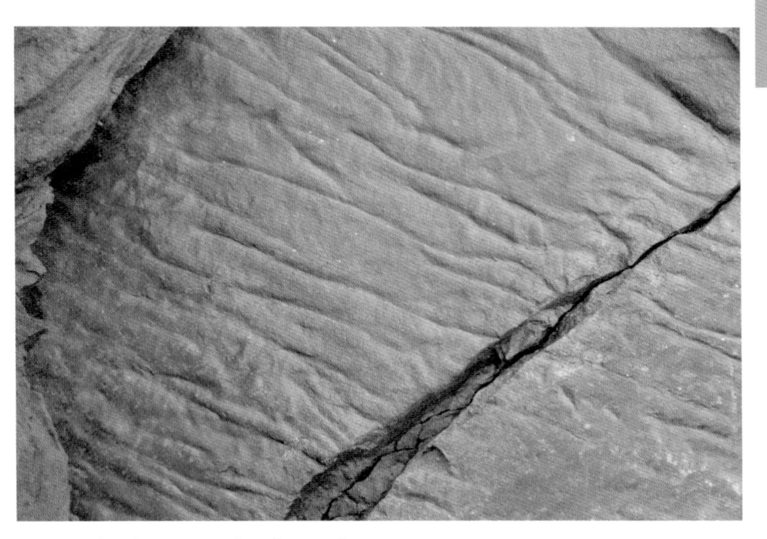

Longitudinal scour marks, Cornwall.

Infilled washout channel, Northumberland.

POST-DEPOSITIONAL STRUCTURES

In this category are those structures that form after deposition of sediment, but usually before it has finally been changed into rock by processes of diagenesis. Different types of sediment, such as sand and mud, deposited one on another may, because of loading, cause disruption of the softer material, resulting in the creation of load and flame structures. Wet, unconsolidated sediment can slump after its initial deposition, and when sediment surfaces dry out, they can develop a pattern of cracks through desiccation.

▼ These very irregular projections of sandstone are load casts, seen as sole marks on the base of a sandstone bed. In this exposure, in North Cornwall, alternating beds of brownish sandstone and grey shale are visible. Load casts form when wet, muddy sediment, which is soft and has the ability to flow, has coarser sand deposited on it. The sand has the ability to sink into the mud, giving rise to these unevenly shaped masses. Load casts have stratigraphic use, because they are found on the undersurface of sandstone beds.

Load casts, Cornwall.

Flame structures and load casts, Cornwall.

▲ Flame structures are closely associated with load casts. In this exposure of Carboniferous rocks, the flame-shaped masses of grey shale were forced up, as wet mud, between coarser, heavier sand lobes, now brownish sandstone load casts. The larger flame is about 4cm (1.6in) tall.

▼ Wet, muddy sediment can become contorted by movement down-slope, producing structures that are reminiscent of folding. In this exposure, which is about 30cm (12in) from base to top, grey mud has slumped among sandy sediment. For sediment to move in this way, no great slope is required, especially when the sediment is waterlogged.

Slump structure, Cornwall.

Slump roll, Sutherland.

▲ This relatively large, rounded structure probably formed as a mass of slumped muddy sediment under water. When seen in cross-section, such structures can resemble folding. A hint of this shape can be seen at the right-hand edge of the roll. The structure is about 60cm (24in) high.

▶ (top) Here, in Charnwood Forest, fine-grained volcanic tuff and mudstone exhibit a sag structure, which is about 15cm (5.9in) deep. In the upper parts, the thin beds can be followed through the feature, but towards its base, they become fragmented. The structure was probably formed because saturated sea-bed sediment was unstable.

▶ Rip-up clasts are formed when a bed of previously deposited clay, shale or mudstone is fragmented by a sea-bed current, often a turbidity current. The moving water is especially erosive if it is charged with sediment in suspension. Clay-rich sediment may be relatively cohesive soon after deposition. Clasts (fragments) can be broken off and deposited after transportation by the water current. In this example, pale clay fragments are surrounded by water-deposited volcanic tuffs.

Sag structure, Leicestershire.

Rip-up clasts, Cumbria.

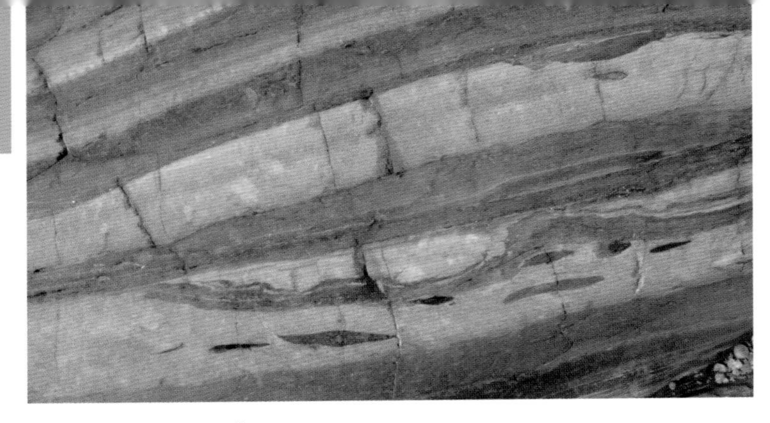

Rip-up clasts, Cornwall.

▲ An exposure of alternating pale grey sandstone and darker siltstone and mudstone is the result of turbidity current deposition during the Upper Carboniferous period. Currents flowing over deposited sediment have torn off flakes of mud, which have been incorporated in the sandstone beds. The largest clasts are about 5cm (2in) long.

▶ (top) When a wet, often muddy, sediment surface lies above water level, it tends to dry out. This process of desiccation, which causes shrinkage and cracking, begins on the surface and works down into the sediment. The areas between the cracks may range in shape from hexagonal, as shown here, to triangular and polygonal. Desiccation cracks usually taper vertically into the sediment, and are commonly infilled by material deposited after their formation, or, as in this example, by minerals. Here, the cracks have been preserved by quartz. In sedimentary rocks, the structures can be found on bedding planes, but may sometimes be observed on the underside of beds as positive features representing the infilling of the cracks. Desiccation cracks are commonly seen on the surfaces of sediment deposited in lakes, shallow marine environments and river floodplains. The hexagonal shape on this surface is 3cm (1.2in) across.

▶ Various sedimentary structures are exposed here on bedding planes of Torridonian mudstone and sandstone. The centre of the exposure shows desiccation cracks, preserved as upstanding, positive features. These are the infillings of cracks formed on a bedding plane that has been removed by erosion. Their shape is not very regular, being polygonal rather than hexagonal. The maximum distance between the

Desiccation cracks, Yorkshire.

Desiccation cracks and ripple marks, Sutherland.

cracks is about 30cm (12in). On an underlying bedding plane, near the base of the picture, there are further desiccation cracks and ripple marks. The ripple marks would have formed first, while the sediment was under moving water, and then, following a lowering of the water level, the desiccation cracks developed, as is shown by the cross-cutting relationship between the two structures.

Desiccation cracks, Orkney.

▲ These desiccation cracks, in mudstone of Devonian age, have been preserved by infilling of coarser, sandy sediment. The maximum distance between the cracks is about 15cm (5.9in). On the bedding plane surface there are other features typical of the drying out of wet sediment. Especially noticeable are the small, curved markings caused by flaking of the drying mud.

▼ Here, a coastal pond has dried out, and the muddy sediment that accumulated in it has cracked into a variety of shapes, ranging from triangles to hexagons and squares. Many of the structures have subsidiary cracks across their surfaces. Because of the very recent nature of this desiccation, no sediment has yet infilled the cracks, but some crumbling has occurred around the margins and fragments have fallen in. A characteristic feature of these structures is that the sediment surface may curl up on drying. This is especially seen towards the top of the picture.

Modern desiccation cracks, Norfolk.

STRUCTURES ASSOCIATED WITH DIAGENESIS

Sedimentary rocks are frequently altered chemically, and these chemical changes produce a variety of structures. This alteration can occur on quite a small scale and may take place while the sediment is changing into rock. The chemicals concerned can be in the sediment, contained in water held in pores between the grains, or may be introduced by water circulating through the sediment.

CONCRETIONS AND NODULES

Different sedimentary rocks can contain discrete structural rock masses, which stand out among the strata. These concretions are often rounded, though they may take on many irregular shapes. The smaller, rounded concretions are often called nodules. Concretions can be many metres in diameter, but are commonly a few centimetres in size. Many follow the stratification of the rock and are arranged in rows, with gaps between each concretion, but they may be joined in groups, with thinner, rather elongate 'necks' joining them. Some concretions have a chemical composition very like that of their host sediment, but the different nature of their cementation helps them to stand out. In some cases, as in the siliceous flints found in calcareous chalk, the concretions have a different chemistry. Other concretions may be composed of pyrite, calcite, siderite, limonite, dolomite or ankerite, while calcareous concretions can have a tough outer layer of pyrite.

Nodules and concretions are usually far more resistant to the processes of weathering and erosion than the host sedimentary rock. For this reason, concretions stand proud of the strata and occur in weathered out masses on the surface below an exposure. They can also protect small towers of sediment from erosion, occurring as a resistant cap.

The origin of nodules and concretions is open to much discussion, but is generally a matter of chemicals in the sediment becoming separated a short time after deposition. Those formed during diagenesis possibly result from chemical precipitation around a nucleus, such as a small pebble or organic remains. Many nodules and concretions are related to specific situations. Some may follow animal burrows or roots, and many contain fossils. It is thought that where the bedding is seen to curve around a concretion it formed early during diagenetic processes, as the relative hardness of the concretion would cause strata to be curved around it during compaction. If the bedding can be traced through a concretion, this was probably formed at a much later stage.

Many concretions have a network of cracks running through them, dividing the concretion into sections. These are often referred to as septarian concretions, and their internal fractures are infilled with a variety of minerals, often calcite. The development of cracks may be the result of loss of fluids, mainly water, which leads to shrinkage and internal breaking of the concretion.

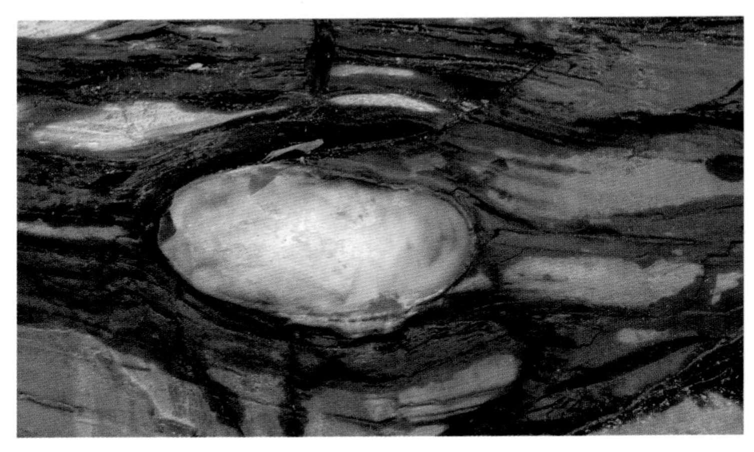

Concretion in shale, North Yorkshire.

▲ This calcareous concretion has shale bedding planes curving around it, suggesting it was formed early during diagenesis, while the sediment was still wet and muddy. A very thin layer of pyrite (fool's gold) can be seen on the concretion's outer surface. The rusty staining on the shale is due to iron-rich water running through the rock. The specimen is about 20cm (7.9in) long.

▶ (top) A 3m (9.8ft) high section of a cliff of Lower Jurassic shale containing numerous concretions is exposed here at Runswick Bay. As is common with the development of many concretions, these have fused together, forming elongated masses. They tend also to follow particular

Concretions, North Yorkshire.

horizons within the strata. These concretions are chemically very similar to the surrounding shale, but have a reddish coating of iron minerals.

▼ The size of concretions can vary considerably. These large, rounded concretions, in Jurassic shale on the North Yorkshire coast, are up to 1m (3.3ft) in diameter. The shale has been eroded by the sea to form a wave-cut platform, with the relatively harder concretions standing up as discrete masses. Though their composition is very like that of the surrounding sedimentary rock, their structure renders them more resistant to erosion.

Large concretions on a wave-cut platform, North Yorkshire.

Ammonite in a nodule, Nepal.

▲ Concretions often form around fossils. This may be because of the different chemical conditions near the dead organism after its rapid burial in muddy sediment. If not buried soon after death, the animal would probably be eaten or decompose. This Jurassic ammonite, from the Kali Gandaki river valley in the Annapurna region of Nepal, has been beautifully preserved in three-dimensional form inside a calcareous nodule. The left-hand half is a cast of the fossil shell, while the right-hand side contains the body fossil. The ammonite is around 5cm (2in) in diameter. Preservation in this way prevents the shell from being crushed by the weight of overlying sedimentary rock.

▶ (top) Certain concretions have a system of internal cracks, which probably develop as water is released from the forming concretion. In this example, an elongated concretion has cracks that have been infilled with calcite. These infilled cracks divide the concretion up into sections with a variety of shapes. The red-brown staining is due to a coating of limonite (hydrous iron oxide).

▶ In some cases, nodules have a concentric internal structure. This nodule, about 5cm (2in) across and formed in relatively coarse sandstone, has a series of concentric layers filled with the iron oxide minerals, yellowish-brown limonite and darker goethite.

Septarian concretion, North Yorkshire.

Ironstone nodule, Yorkshire.

Chert concretion in limestone, Sutherland.

▲ Chert, like flint, may be formed by the hardening of gelatinous silica ooze, which could be derived from the skeletons of radiolarians or sponges on the sea bed. The chert is then incorpoated into marine sediment, especially lime mud, which becomes limestone. Chert forms in discontinuous layers following bedding planes, or as isolated, often rounded, concretions. This chert concretion is about 8cm (3.1in) across and is in limestone exposed on the north coast of Scotland, near Durness.

▶ (top) These steeply inclined chalk strata are picked out by darker, flint-rich horizons, which show regularly spaced, rhythmic deposition. Flint is a nodular, silica-rich material, contrasting in its chemical composition with the calcite-rich chalk in which it occurs, and is usually regarded as a variety of chert. The origin of the silica from which flint is formed may be sea-bed silica, derived from the siliceous spicules of sponges. It is thought that many flints follow the shapes of burrows made in the soft sea-bed sediment by invertebrates. Flint is generally dark-coloured, and, being formed of cryptocrystalline silica, is very hard.

▶ Flint, composed of silica, weathers out from the chalk in which it frequently occurs, because of its hardness and resistance to chemical change. Flint's formation is open to some debate. As with other types of chert,

Flint layers in chalk, Dorset.

Flint concretions, Norfolk.

the source of the flint-producing silica may be organic and derived from sponge spicules, or it may be inorganic in origin. These hard concretions show the characteristic irregular shape of flint and its curved fracture.

Cone-in-cone structure, North Yorkshire.

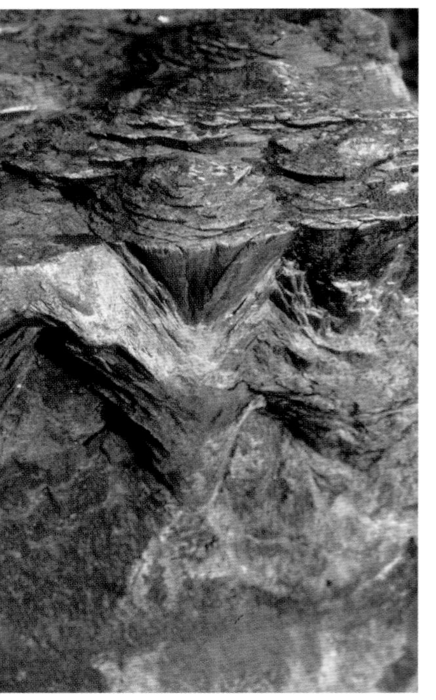

▲ ◀ ▶ Cone-in cone structure consists of a series of small cones set inside each other. It can occur in a variety of sedimentary rocks, including limestone and shale (as seen here), and in or on the surfaces of some concretions. The cones are commonly made of calcite, but the chemistry of the sediment may determine their composition. Pyrite, dolomite, siderite and gypsum have been found in the structures. In some cases, there is a thin clay layer between the cones.

Cone-in-cone is not uncommon, but its origin and mode of formation are not fully understood. It has been suggested that the development of a cone structure is related to a structural change from aragonite to calcite during diagenesis, producing fibrous cones of calcite. This causes pressure within the sediment, and as

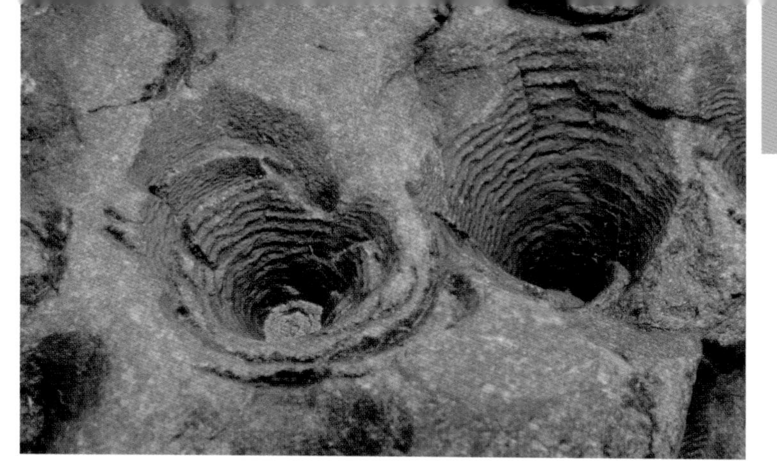

Cone-in-cone, Tyne and Wear.

the cones grow, the shape is related to areas of different pressure. Three pictures are shown, two from Jurassic shales in Yorkshire (one seen from above, the other from the side) and one from Carboniferous strata in Tyne and Wear (seen from above). The diameter of each structure is 2cm (0.8in).

SEDIMENTARY STRUCTURES OF BIOGENIC ORIGIN

Various sedimentary structures are the result of the activity of organisms. These may be found on sedimentary surfaces or as part of the internal structure of a stratum. Some organisms build reefs and mounds of sediment, which then become rock. Traces of the activity of a variety of organisms include infilled burrows made in soft sediment, tracks and trails, and footprints left by larger animals. These all provide important information about how animals lived, and these fossils may provide us with far more information about past life than, for example, an isolated fossil shell. Trace fossils can also be of use as way-up criteria, as many of the organisms that created them lived in burrows made downwards into the sea bed, while others lived on the sediment surface, their tracks and trails indicating the original upper surface of a stratum. Trace fossils are generally given biological names, while the organism that left its trace is given a different name.

Stromatolites, Sutherland.

▲ Stromatolites are laminated mounds of sediment built up by marine microorganisms. Cyanobacteria form mats on the sea bed, which trap sediment and calcite to produce rounded, upwardly convex, limestone structures. These develop by the accumulation of sediment that is trapped and bound together by bacterial filaments. Stromatolites can be a variety of shapes, but commonly, as here, are rounded masses, with very fine internal layering.

Stromatolites are well-known rock-building structures in the Pre-Cambrian and in more recent strata. They were important in the early development of the Earth, as cyanobacteria produced oxygen for the primitive atmosphere. Today, the microorganisms that build stromatolites live in highly saline, shallow marine environments, often around high tide level. Such environments are unfavourable to many organisms, and during the Pre-Cambrian, the development of stromatolites was probably favoured by the lack of creatures that might graze on the cyanobacteria. These stromatolites, of Ordovician age, are exposed near Durness, on the north coast of Scotland. Their fine, laminated structure is clearly seen where erosion has sectioned them. The exposure is about 1m (3.3ft) high.

▶ (top) As well as making discrete calcareous mounds, stromatolites can merge into more continuous structures, as here in dolomitic limestone of Permian age in South Yorkshire, exposed in a cliff about 5m (16ft) high.

Stromatolites, South Yorkshire.

These structures are relatively flat, with slightly convex upper surfaces. This shape indicates that these strata are the correct way up. It is thought that the stromatolites here formed in seawater only a few metres in depth.

▼ These small, calcareous algal mounds occur in Torridonian sediments and are of Pre-Cambrian age. Layers of calcareous material can be clearly seen within the red sandstone that is typical of Torridonian sedimentation. The mounds are 3cm (1.2in) high, and are by far the oldest known fossils in Britain.

Algal mounds, Sutherland.

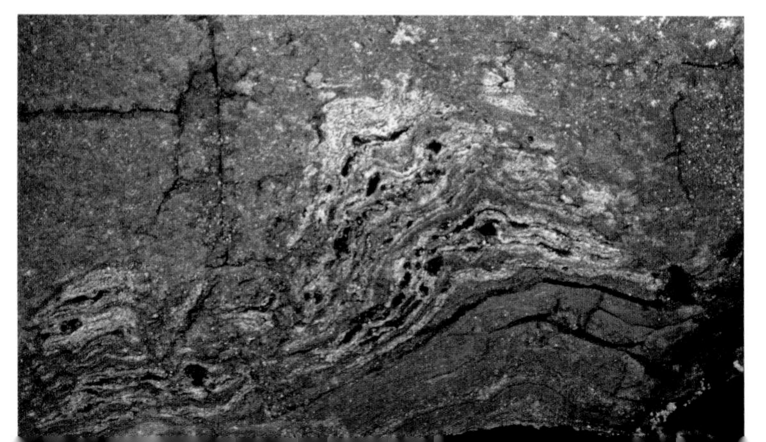

Patch reef, Yorkshire.

▲ Reefs are structures with an upper surface originally close to sea level. Corals are well-known reef builders, but many other organisms, including bryozoans and lime-secreting algae, are known to bind reef sediment and so help to stabilise the structure. Reefs can be elongate structures, forming a barrier between shallow and deeper water, or they may be smaller, isolated patch reefs, which can develop in shallow lagoons. This small patch reef, which has little internal structure, rests on bedded oolitic limestones of Upper Jurassic age. Small mounds of shell sand and lime mud such as this are bound by corals, and may also contain fossils of molluscs and echinoids.

▶ Structures of this type can be found as positive features on a sedimentary rock surface. The original trail or very shallow burrow in the soft sea-bed mud was probably made by a small gastropod, and was infilled with sediment and thus preserved in three-dimensional form. The trails here are about 1cm (0.4in) wide. The first example, named *Scolicia,* is in rocks of Upper Carboniferous age from West Yorkshire, the second in Carboniferous shale in Menorca.

Gastropod trail, West Yorkshire. (below) Gastropod trail, Menorca.

Marine worm burrows, Sutherland.

▲ ◀ These pale, vertical structures are the burrows of marine worms, and they cut through bedding structures in cross-bedded quartzite. From these trace fossils and other evidence, it is believed that the quartzite was deposited as intertidal mud flats and shallow marine sediment. The vertical cliff shows an exposure of about 3m (9.8ft) in height. The upper picture shows the rounded cross-sections of a mass of burrows. *Skolithos* is the name given to vertical burrows of this type.

Dinosaur footprint, North Yorkshire.

▲ This three-toed footprint, about 15cm (5.9in) long, is from Middle Jurassic non-marine sedimentary rocks and was probably made by a small dinosaur. When they are made, footprints disturb the sediment and create trace fossils. The sediment here is fine-grained sandstone, and the structure was formed in relatively damp sand, as the print is clear. If made in very wet sand, the print would have been more indistinct. It is preserved as a cast on the underside of a bedding plane, and the thin ridges radiating from it may be the infillings of desiccation cracks on the sand surface.

FOLDS AND FOLDING

Folded rocks are some of the most striking geological features. These structures may be on a very large scale, measuring many kilometres across, or they can be only millimetres in size. Rocks are bent into folded shapes by lateral compression at depth in the Earth's crust, and folds may form over very long periods of time, whereas faulting can happen quite suddenly. Folding events usually occur during crustal movement caused by plate tectonics, and are generally associated with mountain belts. The theory of plate tectonics suggests that the crust and the higher part of the mantle (the layer below the crust) consist of a series of relatively rigid plates. These are of varying size, six large ones and a number of smaller ones. They are mobile, and where they collide or pull apart, many geological events occur, including volcanic eruptions and mountain building.

As soon as a rock surface is tilted or becomes part of a fold, it acquires two important structural features, dip and strike. The true dip of a sloping rock surface is the greatest angle that can be measured down that surface, using a clinometer. This angle also has a direction. Dip is therefore recorded with two figures: the angle of dip and its compass bearing. Apparent dip is the slope of a dipping rock surface that may be observed, for example, in a cliff face, but may not be the true (maximum) dip angle. A series of strata that appear to be horizontal may, in fact, be dipping quite steeply towards or away from the observer. The direction at right angles to the dip direction is the strike, which is recorded only as a bearing. These two properties of tilted rock surfaces are fundamental to geological mapping and the understanding of rock structures.

Folding of rocks is very much influenced by a number of factors. At relatively high levels in the crust, where pressures and temperatures are not extreme, sedimentary rock surfaces may slide over one another, by a process known as flexural slip. This is rather like the way pages of a paperback book will move against each other if the book is folded. Buckling can occur where a layer of competent rock, such as sandstone or limestone, lies between incompetent beds of, for example, shale or clay. The competent layers respond

Dip and strike, Sutherland.

▲ On the west coast of Sutherland, Torridonian sandstone strata dip westwards to the right, the large exposed bedding plane emphasising this structural feature. The strike of these beds is almost north-south, following the level of the sea on the dipping rock surface.

---

to lateral compression differently from the beds above and below, becoming folded and maintaining their thickness. The wavelength of the folds is dependent on the thickness of the affected layer, the thinner beds producing tighter folds. The incompetent layers are less easily folded and tend to cleave (fracture).

When rocks are subjected to pressure during folding, minerals such as quartz can undergo solution and move through the rock. This type of process helps to explain how relatively brittle rocks, including many sedimentary strata, become mobile and can be folded. Fractures in folded rocks usually contain mineral infillings, commonly of calcite and quartz.

At considerable depth in the Earth's crust, the high temperatures and pressures cause minerals making up the rocks to change by recrystallisation. It is under these conditions that igneous and metamorphic rocks tend to be folded, in an almost plastic state.

Where such conditions occur, regional metamorphism can change rocks completely.

Some folds are formed by other processes in the Earth's crust. Large-scale forceful intrusion of igneous rocks such as granite can cause disruption of the surrounding country rock. Sedimentary strata may be arched above the intrusion and so dip away from it. Strata around the sides of the intrusion can be compressed and folded. Rock salt and other evaporite rocks have a low density, and flow upwards under pressure from overlying strata forming a salt dome, which affects surrounding rocks in a similar way to a forceful igneous intrusion, but this occurs at a relatively high level in the crust.

FOLD SHAPES

There are two basic terms used to describe fold structures. Fold shapes can vary considerably, and many types are possible. An antiform is an 'upfolded' shape, resembling an arch, whereas a synform is the opposite, a 'downfolded' structure. These terms are related closely to the words anticline and syncline. An anticline is an antiform in which the older rocks are in the core of the fold. A synclinal fold is a synform with the younger rocks in the core. In order to establish in the field the true shape of a fold, the relative ages of the folded rock layers must be established. If, for example, an apparent syncline has the older rocks at its core, it is shown to be an overturned anticline, and is described as a synformal anticline. Basins and domes are characterised by strata that dip in all directions from a central area. Basins are synclinal and domes anticlinal. Periclines are similar structures, but are elongated. Basins, domes and periclines are relatively large structures.

When describing fold geometry, a number of terms are used. The hinge, or closure, of a fold is the area where bending is greatest. On either side of the hinge are the limbs. The hinge may be relatively sharp, with fairly straight limbs, and a hinge line is then recognised. Where the hinge is less precise, and the limbs more curved, there is said to be a hinge zone. The limbs in a series of folds are the areas between the hinges, and each limb is part of two juxtaposed folds. The axial plane lies between the two limbs of a fold, at equal distance from each and bisecting the angle between them. It is an imaginary surface only. The term 'axial plane' is used when considering a single folded layer of rock. When the fold involves multiple layers, the term 'axial surface' is used for the theoretical surface linking the hinge lines of the various layers. The fold axis is the line where the hinge zone is cut through by the axial plane. The angle of plunge is that measured between a dipping hinge and the horizontal.

The dip of the axial plane establishes the inclination of a fold. Where the axial plane is more or less vertical and the limbs dip at similar angles, in opposite directions, the fold is described as upright, or symmetrical. With an asymmetrical fold, the limbs dip at different angles to each other and the axial plane is not vertical, but dipping at a lower angle, giving the alternative name, an inclined fold. Where one limb of an inclined fold is overturned, the fold is termed an overfold. In extreme cases, when the axial plane is more or less horizontal, the fold is known as recumbent.

There are a number of types of fold profile. To identify the profile, a fold must be viewed at right angles to the axis. Each of the beds forming the limbs of a parallel fold has a constant thickness when measured at right angles to the top and bottom surfaces of a layer. In folds called similar folds, the thickness of each layer is constant when measured parallel to the axial plane. Chevron folds are characterised by long, straight limbs, with sharp, angular hinges between them. They typically occur where competent layers of rocks such as sandstone are interbedded with incompetent layers, often of shale. Flexural slip can take place between these different layers. The folds may form as a series, producing a striking, zigzag structure. Where such folding takes place on a very small scale, the folds are known as kink bands. As with chevron folds, the limbs are straight and the hinges sharp. Kink bands are produced in finely laminated metamorphic rocks, typically slates.

When folds and tilted strata are eroded, landforms such as escarpments can form. Here a number of beds having the same dip produce a series of dip slopes with their eroded edges forming steep escarpments. Depending on the resistance of the various layers within the dipping rock sequence, some of the escarpments may be more prominent.

Dip and strike, Somerset.

▲ The dip and strike of these alternating Lower Jurassic limestone and shale beds is well displayed here on the Somerset coast. The dip is shown by the obvious slope of the more resistant limestone strata, which dip landwards. The less resistant shale beds have been eroded, and form shallow depressions between the limestones. The strike direction can be traced along each bed into the distance. Older beds can be seen to dip below progressively younger beds towards the land. The effect of marine erosion on these strata has been to create a series of small-scale escarpments and corresponding dip slopes.

▶ Here, a relatively symmetrical anticline, with each limb dipping at much the same angle, has been eroded in a sea cliff. The alternating sandstones and shales are of Carboniferous age, the resistant sandstones indicating the fold structure where the less resistant shales have been eroded more readily. The dip observed here is near to the true dip, and the strike direction is shown by the grey-coloured rib of rock coming from the anticline to the bottom left of the picture. Erosion has removed rock from the top of the fold, and slumping of rock debris obscures part of the centre of the anticline.

Anticline, Cornwall.

Open anticline, Devon.

▲ On the north coasts of Cornwall and Devon, folding is commonly extreme, with very tightly folded anticlines and synclines in the Carboniferous shales and greywackes. Here, however, is a very open anticlinal fold, which has been eroded into a 'barrel-vaulted' cave. The beds are folded into a gentle arch shape and their thickness, measured at right angles to the base and top surface of each bed, remains constant. This fold can therefore be termed a parallel fold.

▶ (top) This tight, angular anticline has formed in alternating greywacke sandstones and shales. There has been movement of the sandstone beds against the shales, allowing the fold to form. In places, small faults have developed. The axial surface of the fold is curved, and leans to the left at the top of the picture.

▶ This rounded anticlinal fold, well exposed on the Cornish coast, is made of Carboniferous sandstones and shales, which formed from deposition by turbidity current flows. The limbs of the fold dip away from the hinge zone along the length of the pericline. They would also have dipped away from the hinge at both ends of the structure, but the near end has been removed by erosion. The sandstones are cut by numerous joints resulting from tension during folding. This anticlinal pericline has a shape rather like an upturned boat.

Tight anticline, Devon.

(below) Pericline, Cornwall.

Dome structure, North Yorkshire.

▲ These Lower Jurassic shales are well exposed at low spring tide on the wave-cut platform at Robin Hood's Bay. The dip of the strata changes around the bay, but is always towards the shore. Miniature scarp and dip slopes have been formed, with the steeper scarp slopes, here in shadow, facing out to sea. The strike, the direction at right angles to the dip direction, is shown by the curving tops of the miniature escarpments. These curve because the dip changes around the bay and the strike maintains its relationship to the dip direction. The beds become increasingly younger towards the shore. The overall structure here is a dome, with strata dipping away from its centre, which is offshore out in the bay. The shape of the coastline and the broad sweeping curve of the bay generally follow the strike of the rocks and are related to the structure.

▶ (top) Here an asymmetrical syncline is seen in sandstones and shales of Carboniferous age. The left-hand limb is overturned, while the right-hand one dips at a relatively low angle. The strata in this overturned fold have been disrupted and faulted, and the softer shale beds have been removed by erosion in places. This cliff face is about 3m (9.8ft) high.

Overturned syncline, Cornwall.

Folding, Dorset.

▲ Alternating limestones and shales of the Jurassic Purbeck Beds have been compressed into remarkable folds. Some of the limbs dip very steeply. The anticline on the left shares a limb with the syncline to its right. The limestones are more brittle than the shale beds and have fractured in many places. This is part of the well-known 'Lulworth Crumple'.

Overfolded anticline, Cornwall.

▲ This overfolded anticline is clearly seen in the lower left-hand quarter of the picture. The left-hand limb is more or less horizontal, while the right-hand limb has passed through the vertical and is now overturned, suggesting more compression from the left. The rocks involved in this folding are Carboniferous sandstones and shales. The section is around 3m (9.8ft) high.

▶ Two examples of recumbent folds are shown. In the top picture, from Millook Haven, the lower limb has been overturned to such an extent that the fold is recumbent. A prominent, more resistant bed of sandstone has been much fractured during folding, the fractures being infilled with pale quartz. These rocks are of Carboniferous age. The exposure shown is about 3m (9.8ft) high.

Upper Devonian slate and greywacke form a recumbent fold in the lower picture from Jangye Ryn on the Lizard Peninsula. This area was visited in 1822 by Adam Sedgwick, a pioneering geologist, and he found such a structure 'difficult to account for by the action of mere mechanical forces'.

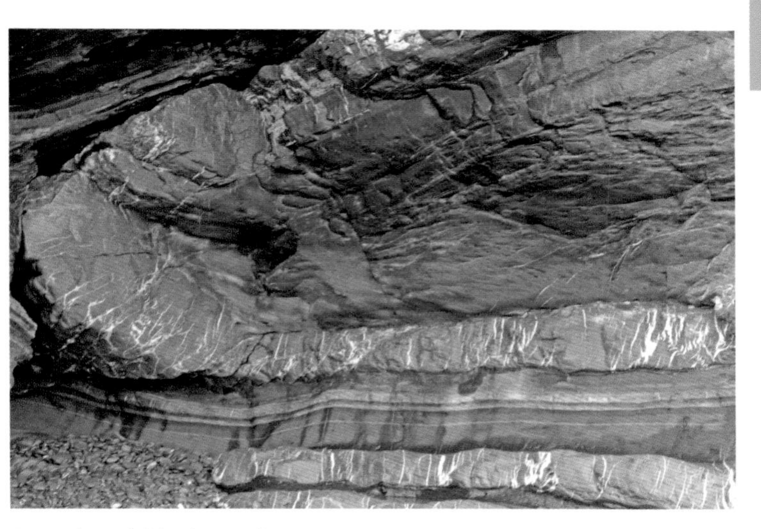

Recumbent folds, Cornwall.

Buckling, Cornwall.

▲ The Mylor Beds that are exposed here are grey to dark blue slates, containing coarser bands of sandy sediment and numerous quartz veins. These rocks are of Devonian age. Very tight folding on a small scale is clearly seen in the narrow, highly buckled quartz veins. The narrower veins have been folded more tightly than the broader veins. Although lateral compression causes buckling of the quartz veins, the surrounding slate layers are not affected. There is evidence for two periods of compression of these quartz veins, as very small folds are superimposed on the larger structures. The area shown is about 60cm (24in) across.

▶ (top) Chevron folds are characterised by limbs that are relatively long (not necessarily all the same length) and angular hinges. This famous exposure at Millook Haven shows alternating siltstone and shale beds, overfolded to give a zigzag effect. The tight folding in the fold hinges is created by the movement of the less competent shale beds between the more durable siltstone layers. In many parts of this cliff, small-scale faulting has occurred, especially in the fold hinges, and the siltstone beds are fractured by tension gashes filled with quartz. The whole cliff is about 35m (115ft) high.

▶ Dark slate exposed on the north coast of Cornwall, at Boscastle, has been compressed to produce a small-scale fold. Tougher beds within the slate pick out the folding, which appears as a small kink or flexure in the rock. The limbs of the fold are typically straight, and the folding is confined to a small area. The thrift plants give the scale.

Chevron folds, Cornwall. (below) Kink fold, Cornwall.

Kink folds, Cornwall.

▲ Here, a series of kink folds is exposed on a low cliff. Although the limbs in such folds are usually straight, some in this example are slightly curved. These folds are very similar to chevron folds, but on a much smaller scale. The field of view is about 60cm (24in) from top to bottom.

▶ (top) The dark greenish structure is a small dyke that has been folded within a mass of pale, acid Lewisian gneiss. The dyke has a basic to ultrabasic composition. Such folding of what is a structurally competent feature has taken place at great depth, when temperatures and pressures were extreme, producing high-grade regional metamorphism. The dark-coloured dyke is about 1m (3.3ft) across.

▶ High-grade metamorphic rocks frequently show tight folding, which has occurred when the rock has been in a plastic state. Here, the characteristic light and dark banded structure of the gneiss emphasises the folds. The field of view is about 60cm (24in) across.

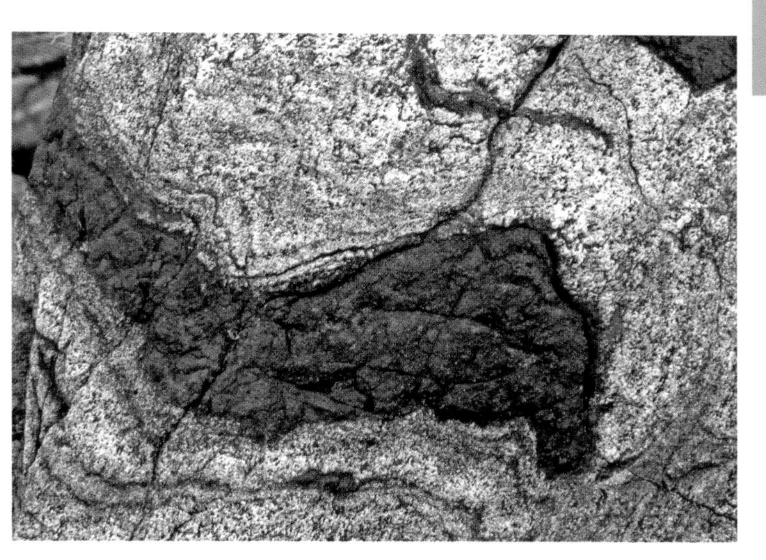

Folded dyke, Sutherland.

Folded gneiss, Sutherland.

STRUCTURES RELATED TO TENSION

FOLD MULLIONS

Fold mullions are a somewhat enigmatic structure, not uncommon in areas of considerable rock deformation. They are large-scale, columnar structures named after the carved stone window frames often found in old buildings. Mullion structure looks something like cut tree trunks stacked on top of each other. The structure is associated with both folding and stretching. Mullions generally occur where competent and incompetent layers of rock come into contact, and are formed by the folding of bedding and cleavage planes, the structure then undergoing stretching parallel to the cylindrical mullions, which may be, in effect, large-scale rodded structures.

▼ This structure, on a smaller scale than fold mullions, is nevertheless related to mullions, and the result of similar rock deformation and stretching. Rodding is found in highly deformed rocks. The quartz comprising the structure may be an original component of the rock, or may have been introduced at a later date, possibly during deformation. In some cases, the rods are formed by the stretching of quartz pebbles. Small centimetre-sized pebbles may be stretched out into rods measured in metres. In this example, a mass of small, parallel quartz rods dips into a mossy rock surface, looking rather like crystals of satin spar gypsum.

Rodded quartz, Sutherland.

Mullion structures, Sutherland.

▲ Characteristic fold mullions, with longitudinal grooves and rounded, columnar outlines, are seen here at Coldbackie, on the north coast of Scotland. The rock is a fine sandstone, with mica occurring on the surfaces of some of the mullions.

BOUDINAGE

Boudinage structures may be produced under tension when competent rock is surrounded by less competent material. As stretching occurs, competent layers within the rock mass may become broken and pulled apart, often with narrow necks between the fragments (boudins). These are often rounded and lens- or pillow-shaped. This process is facilitated by the less competent surrounding rocks, which allow the stretching of the competent layers of rock, curving around the boudins. When seen in three dimensions, boudins have an elongated, almost columnar structure. The geological term 'boudin' is derived from the French boudin, meaning a type of sausage. Boudins have been called 'pull-apart' structures.

Boudinage, Sutherland.

▲ In the example where a hand lens is used for scale, a competent pale quartz and feldspar layer has been stretched and partially broken among less competent dark rock. Two of the boudins are linked by a relatively thick quartz neck. It can be seen how the dark rock is curved around the quartz and feldspar boudins.

▲ The above example has a coin to show the scale. These quartz-rich boudins are small and linked by thin necks. The resemblance to a string of sausages is striking. The dark surrounding rock is of ultrabasic composition.

▼ Here, boudinage is seen on a much larger scale, as indicated by the Ordnance Survey map. In the foreground, there are two large, pale, acidic boudins, linked by a thinner neck. A series of smaller boudins, of similar composition, is visible on the sloping ground above the map.

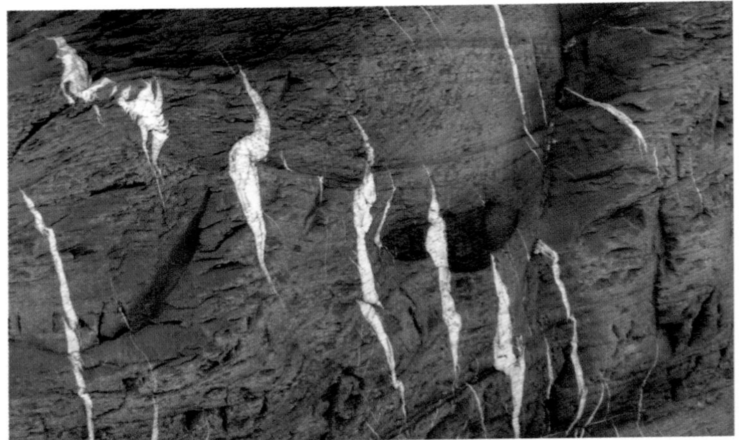

Tension gashes, Cornwall.

▲ Tension gashes originate from small fractures caused by stretching of rocks. These fractures are then infilled by minerals such as quartz, as here. Tension gashes are widest in the middle, decreasing in thickness towards each end. They often occur in arrays, with many forming together. When the fractures are parallel to each other, they are known as 'en echelon' tension gashes. S- or Z-shaped examples are referred to as sigmoidal tension gashes. These shapes may result from the rotational shearing of the rock continuing after the initial fracture has occurred. The largest gashes here are about 15cm (5.9in) long.

FAULTS AND FAULTING

A fault is a fracture in the rocks of the Earth's crust along which there is relative movement of the rocks on opposite sides. A fault may move often in irregular jerks or by continuous creeping. Fault movement is the main cause of earthquakes. The surface along which the movement occurs, the fracture plane, is called the fault plane. The term fault surface is often used, as it may not be a smooth, neat, plane surface. This surface can have scratches (slickensides) and marks on it, caused by the rubbing of rock masses during faulting.

It is not always easy to determine the actual movement of a fault. There are both vertical and horizontal components in most fault movements, but slickensides will give a good indication of the displacement that has occurred. Faulting may move rocks only a few centimetres, or many kilometres, and can cover a wide geographic zone, with a number of parallel faults involved.

The mass of rocks above the fault plane, when this is not vertical, is referred to as the 'hanging wall', and the block below the fault plane as the 'footwall'. The terms 'downthrow block' and 'upthrow block' are also of use, as they indicate the relative movement on a fault. A sloping fault plane has the property of dip from the horizontal, just as a sedimentary bedding plane does. Similarly, a fault plane has strike, the compass direction at right angles to the direction of its dip. Hade is the angle made between the fault plane and the vertical. If the fault plane dips at an angle greater than about 45°, it is said to be a high-angle fault, and below 45° the fault is of low angle. The throw of a fault is the vertical displacement on the fault plane. This can be measured with accuracy by following the same bed or horizon on each side of the fault. The term 'heave' is applied to any horizontal movement along a fault plane.

Rocks on either side of the fault plane are often fragmented during movement, and fault breccia may be formed. This is a mass of crushed and broken rock, often of mixed grain size. At times, fault movement produces a crushed, clay-like material referred to as gouge. In many cases of deep, large-scale faulting, a crushed and fine-gained rock called mylonite is created. Because this is an altered rock, it is usually classified with the metamorphic rocks.

Landforms associated with faulting are often related to the juxtaposition of different rock types on either side of the fault plane. Escarpments and cliffs commonly form in these places and may follow the line of the fault. The more easily eroded rock on one side of a fault will generally form the lower terrain, and the more resistant rock the high ground. Because the line of a fault is often quite straight, any cliff or escarpment produced is likely also to be straight.

A number of fault types are recognised and classified according to the direction of their movement and the angle of the fault plane.

NORMAL FAULTS

A normal fault is a type of dip-slip fault. The main movement of a normal fault is parallel to the dip of the fault plane, and is such that the hanging wall moves down relative to the footwall. These faults are probably the most common, and take place when the Earth's crust is stretched and extended. The fault plane generally dips at a high angle, usually more than 45°, but if the fault plane of a dip-slip fault is at a very low angle, the fault is called a lag fault. Many normal faults have a vertical displacement (throw) of only a few metres or less; however, some have far greater movement.

When a number of parallel normal faults are found together, downthrow blocks called 'rifts' or 'graben' may occur between them. A relative upthrown area between two faults is called a 'horst'. Step faulting is the name given to a series of parallel normal faults, each with downthrow to the same side, producing stepped blocks of displaced material.

▶ (top) A normal fault cuts this exposure of gneiss. The fault movement is shown by the displacement of the two dark bands. The field of view is only 15cm (5.9in) across, and is useful in showing how distinctive layers on either side of the fault plane can be matched up to indicate the fault movement. In this case the throw is about 1cm (0.4in).

▶ Where there is a horizon or stratum that can be easily recognised in the section showing a fault, and this stratum occurs on both sides of the fault, it is relatively easy to see the throw (vertical movement) of the dip-slip fault. The pale layer of sandstone is seen to have moved down

Small-scale normal fault.

Normal fault, Menorca.

the fault plane to the left-hand side. This side, lying above the fault plane, is the hanging wall, and is on the downthrow side. The vertical displacement is about 1m (3.3ft). The fault plane is only seen in section, and appears as a slight groove in the cliff face. As is common with such structures, this is a point of weakness. Where rock has been ground up by movement of the fault, water running down the cliff has eroded the fault plane. The rocks are of Triassic age.

Fault plane, Somerset.

▲ Fault planes are often clean rock surfaces and when the rocks of the hanging wall, above the fault plane, have been removed by erosion, as in this example, their features can be readily observed. A number of faults can be seen in the nearby cliff face. Though this fault plane and those in the cliff are at relatively low angles, they are dip-slip, normal faults. The left-hand fault plane exposed in the cliff face, just beyond the figures on the shore, has a prominent pale limestone stratum indicating the downthrow side to the left. The fault plane is a flat surface with few irregularities, striking away from the camera viewpoint. Close observation will show that it is covered with many markings and grooves running down the surface parallel to its dip. These may well be a result of movement on the fault. The rocks here are of Lower Jurassic age.

▶ Evidence for faulting can be of various types. When two distinct rocks are found close together, faulting may be suggested. Here, greenish marl and grey shale to the left (on the downthrow side) and red-coloured marls to the right have been involved in faulting. Other evidence for faulting in this exposure is the gully running down the cliff face. This has been produced by erosion, caused by running water, an important mechanism of cliff erosion. Products of this erosion obscure the fault plane and the exact relationship between the two different rocks.

Erosion of fault plane, Somerset.

Normal fault, Northumberland.

▲ The Howick Fault, which is exposed on the coast of Northumberland, has associated with it a complex of a number of small to medium-sized faults. Here, a near vertical cliff section shows a normal, dip-slip fault in Carboniferous sedimentary rocks, mainly sandstones and shales. It is relatively easy to detect the movement and the amount of displacement from a careful study of the strata exposed. The downthrow can be seen to be on the right-hand side of the dipping fault plane. It is not possible in this section to say what, if any, horizontal movement has occurred. The section shown is about 2m (6.6ft) high, and the small, round holes in certain strata are where a geologist has taken rock specimens with power tools.

▶ (top) Often a number of faults occur in association. Here, two normal, dip-slip faults are exposed in a cliff, with downthrow between them. This has produced a small-scale rift, or graben structure, with a downthrow block between the two fault planes. Such faulting causes extension of the Earth's crust, if here only on a small scale, as the downthrow block has, in effect, slipped from a higher stratigraphic level between the two fault planes. A number of tension joints are visible in the downthrow block. These are infilled with white quartz. The section is about 2m (6.6ft) high, and shows rocks of Carboniferous age.

▶ The escarpment of Almannagjá is in the Thingvellir National Park in southern Iceland. It follows the strike of a large-scale normal fault, affecting an active fissure volcano of the Hengill volcanic system, with the eastern face downthrown by up to 40m (130ft). This normal fault is

Normal faults, Cornwall.

viewed here from the north-east, and the line of the escarpment is seen to be consistently straight. The Almannagjá Fault is on the north-western side of a down-faulted graben, or rift structure, with the low-lying Thingvellir Plain to the south-east.

Iceland lies on the Mid-Atlantic Ridge, where the European tectonic plate and the North American plate are moving apart. It is for this reason that Iceland is such a volcanically active region. This rift system is on the plate boundary, and movement from tensional stresses occurs in bursts, causing earthquake activity and volcanism. In 1789, the Thingvellir Plain subsided more than 60cm (24in), with rifting taking place, accompanied by flooding.

Normal fault, Iceland.

Evidence for faulting, Menorca.

▲ At Cala Morell, on the north coast of Menorca, in the Balearic Islands, there is a well-known fault that brings Miocene conglomerates and sandstones against dolomitic limestones of Jurassic age. Here, evidence for faulting is shown. To the left-hand side (west) of the bay is an extensive outcrop of Miocene conglomerates and boulder beds, and on the opposite side (east) the Jurassic limestones are exposed. Both rocks are now at about the same stratigraphic level, whereas, when formed, the Miocene conglomerates would have been stratigraphically much higher. Such proximity of rocks of different ages is not on its own proof of faulting, but faulting is a likely explanation. The fault plane is not visible here, because it is under the sea, but it is exposed on the nearby coast.

▶ The Cala Morell fault is well exposed as a normal fault with an almost vertical fault plane. The rocks on both sides of the fault are strikingly different. To the right are reddish-brown conglomerates of Miocene age, while to the other side are pale grey dolomitic limestones, formed during the Jurassic period. The downthrow side is where the younger rocks are, to the right of the picture. A careful study of this fault has been made by Spanish geologists, and it is thought that the fault moved at least once. The fault plane has been referred to as a 'zone of confusion'. The more

Faulting, Menorca.

recent movement of the fault may well have caused the contortions in the Miocene strata seen to the right of the fault, beyond which are other small normal faults.

Low-angle normal fault, (lag fault), Cornwall.

▲ Where extensive folding occurs, as here in North Cornwall, there is often associated faulting. These rocks are greywackes, sandstones and shales of Carboniferous age. The angle of this fault plane is less than 45°, and a close study of the folded rocks involved shows the direction and amount of movement. Some of the beds, notably a brownish sandstone near the left-hand margin of the picture, are hardly broken, but bend at the fault. Other beds, such as the thin sandstones with alternating grey shales near the centre of the picture (where a group of plants of rock samphire are growing), are affected and broken by the fault movement. These beds have only been displaced less than 1m (3.3ft) to the left. Below the fault plane, a number of sandstone beds have been bent by dragging to the left, indicating the direction of fault movement. Many of the beds have been affected by tension, with tension-produced joints being infilled with pale quartz.

REVERSE FAULTS

Reverse faults, like normal faults, are dip-slip faults. The difference lies in the direction of movement parallel to the fault plane. Reverse faults are caused by compression of the rocks in the Earth's crust, causing crustal shortening, and so the movement of this type of fault is of the hanging wall moving up the fault plane relative to the footwall. Commonly the fault plane is at a high angle, but if this type of movement occurs on a very low-angle fault plane, the term 'thrust fault' is used.

Reverse fault, Cumbria.

▲ This very small-scale reverse fault affects fine-grained volcanic ash of Ordovician age. Compression has moved the hanging wall up the low-angled fault plane from left to right. It is possible to match up the beds of ash on each side of the fault to see that very little movement has occurred. The small scale of this fault should be compared with the huge thrust faults shown in the next images.

THRUST FAULTS

Thrust faults, as has been mentioned, are reverse faults with very low-angled fault planes. The fault plane generally dips at far less than 45°, and may be more or less horizontal. A thrust fault is a dip-slip fault, and in some cases the movement is of the order of many kilometres. The hanging wall is termed the 'thrust sheet', and this mass of relatively older rock may be carried great distances from depth over relatively younger rocks. Thrust faults therefore have great geological significance, in that they can move huge masses of rock from mountain roots to higher levels in the crust. This crustal shortening, often on a large scale, is accommodated by increasing the thickness of the crust. In the great mountain chains of the Earth, such as the Rockies, Himalayas and Alps, thrust fault horizontal movement can be measured in hundreds of kilometres.

The very well researched Moine Thrust Zone in north-west Scotland may have a similar horizontal component. Here, as with many thrust zones, there are a number of successive thrust faults in a wide area. These provide a series of 'slices' of rock, separated by thrust faults, and these 'slices' are brought one above another towards the surface from great depth, and move over and against a resistant mass called the 'foreland'. The rocks brought from depth are generally very different from those of the foreland, and are high-grade metamorphic rocks, which may have been carried as far as 75km (47 miles). Dynamic metamorphism, brought about by the crushing and stretching of rock on large-scale thrust planes, creates the rock called mylonite.

▶ Large-scale thrusting occurs in mountain chains, when huge areas of compressional movement force great masses of rock over one another. The Glencoul Thrust, exposed so magnificently here on the far shore of Loch Glencoul, is one of a number of thrust faults in this area that were active during the Caledonian orogeny. These thrusts moved from the east, the right of the picture, against a resistant foreland to the west. Each of the thrusts carried a 'slice' of older rock from great depth and moved it on to younger rocks.

From sea level upwards is exposed the grey, hummocky Lewisian gneiss. On top of this lie stratified Cambrian and Ordovician sedimentary rocks. Above these Cambro–Ordovician sediments is the thrust plane. This makes a definite step on the profile of the skyline, and separates the sedimentary rocks from another mass of irregular grey Lewisian gneiss above. The oldest rock in the area is thus repeated here as a result of thrust movement, and lies over much younger rocks, where it has been thrust up the fault plane. Movement on this and other thrusts in the area could have brought rocks from great depth a distance of 75km (47 miles) from the east.

▶ A hand specimen of mylonite from the Moine Thrust is shown here, with alternating light and dark bands of minerals. The rock is mainly composed of fine- to medium-grained rock flour formed by pulverisation during movement on the thrust plane.

Thrust fault, Sutherland.

Mylonite, Wester Ross.

Thrust plane, Sutherland.

▲ (above and right) This thrust plane, which carries metamorphic rocks (mainly schist and related rocks) of Pre-Cambrian age over Cambro–Ordovician sedimentary rocks, is the most easterly of the series of thrusts affecting the geology of north-west Scotland. It lies above the Glencoul Thrust. Following the discovery by Charles Lapworth of the Moine Thrust in the 1880s, the area was surveyed in detail by officers of the Geological Survey, including the famous geologists Ben Peach and John Horne, late in the 19th century and early in the 20th. The area became a world-renowned classic for the study of tectonics and mountain building.

In these pictures, the Moine Thrust plane can be recognised as a sharp, clean surface separating pale, younger rocks below from the darker Pre-Cambrian Moine rocks above. Immediately overlying the thrust plane, the rocks have been crushed and streaked out by the thrust movement, forming a rock called mylonite. This large-scale thrust faulting took place during the Devonian period, towards the end of the Caledonian mountain building events. The thrust outcrops for about 190km (120 miles), from the north coast of Scotland to the Isle of Skye, and is exposed for much of this distance, forming an escarpment in many places.

▶ This thrust plane affects Cambro–Ordovician limestones. The cliff to the left is of rock that has been thrust up the fault plane over the limestones on the right. The view is looking north. Landforms controlled by geological structure and associated with thrusting can be appreciated here. The cliff to the left has been eroded back and now forms the steep side of a river

valley, contrasting with the less steep slope on the right (east), which is the thrust plane. As this is limestone country, the river only occupies this valley at times of high run-off. There are numerous sinkholes and enlarged joint systems in the limestone, into which surface water flows.

Thrust plane and landforms, Sutherland.

Thrust fault, Menorca.

▲ Faulting can be on a vast scale, as with the Moine and associated thrusts, or on quite a small scale, as here. This low-angle thrust plane separates folded and contorted Carboniferous shales and thin sandstones below from less folded Carboniferous shales above. The section is about 1m (3.3ft) in height. Some disturbance of rock can be seen next to the fault plane. Small-scale thrusting and displacement like this is not uncommon in association with folding. White quartz veins can be seen in the rocks overlying the fault plane where joints have developed. As with the extremely large-scale thrusts previously described, this thrust has a very well defined plane.

STRIKE-SLIP FAULTS

Faults that have movement parallel to the strike of the fault plane are called strike-slip faults. They are thus different from normal, reverse and thrust faults, where movement is parallel to the fault plane's dip. The terms 'wrench' or 'transcurrent' fault are sometimes used for strike-slip faults. Such faults are less common than dip-slip faults, and often have a fault plane that is vertical or very nearly so. Usually there is very little vertical movement, virtually all the displacement being lateral. The movement is referred to as sinistral (left handed) or dextral (right handed). This displacement can be determined by looking across the fault and working out the direction of movement of the block on the far side of the fault plane. If it has been displaced to the left then it has sinistral movement.

JOINTS

A joint is a fracture in rock along which there is no movement. Joints vary from small cracks and fractures to large-scale structures, and may be both horizontal and vertical, set mainly at right angles to each other. Different joint structures, with varying origins, are found in the different rock types, and are dealt with in the relevant rock chapters. Joints are frequently important in landform development, as they provide ideal passages for fluids from the surface. These cause weathering and erosion, which in turn can create landforms of various types.

▼ When movement takes place along a fault plane, the rocks on either side of the fault rub and scratch against each other, especially when there are irregularities. The rock surfaces may be grooved by such contact, or, in some cases, polished and smoothed. Here, slickensides (scratches) on a fault plane have been slightly mineralised by fibrous calcite. Slickensides are of considerable geological interest, as, because they tend to be parallel to the fault movement, they indicate the actual direction of movement on the fault plane. The scale of the pictures can be deduced from the limpet shell on the rock surface.

Slickensides, Cornwall.

Fault breccia, Menorca.

▲ In many cases, the rocks involved in faulting become broken at and near the fault plane. The broken rock, which can be involved in further fault movement, becomes jumbled, and breccia, a rock composed of irregular, angular fragments, is created. Here, on the north coast of Menorca, limestone of Jurassic age has been faulted and breccia created near the fault plane. The angular limestone fragments, the largest about 3cm (1.2in) in size, are cemented by an iron-rich clay.

UNCONFORMITIES

An unconformity is a geological structure representing a break in the geological sequence, and such breaks, or 'time gaps', can prove both fascinating and frustrating to geologists. Many millions of years of the Earth's geological history may be unrepresented at an unconformity, and it can be a matter of speculation as to what events occurred at the location where the unconformity is found. The time gap may indicate a major period of erosion or a time of non-deposition.

A number of different types of unconformity are recognised. An angular unconformity has eroded, folded or tilted rocks beneath the plane of unconformity and less deformed, often horizontal strata above it. The term nonconformity is used if an igneous rock or metamorphic rock is immediately below the unconformity and sediments lie above this rock. When only sedimentary rocks are involved, there are two terms used. A paraconformity is where very similar sedimentary strata lie above and below the unconformity, whereas in a disconformity, the sedimentary rocks above and below the plane of unconformity are different.

The passing of geological time and the occurrence of geological events are recorded in rocks and their structural features. The vast thickness of sedimentary rocks is a record of deposition, mainly on the sea bed, but also, to some extent, on the land surface. Subsequent processes may fold and break these layers and uplift them to form mountain ranges. Igneous intrusions and volcanoes are evidence of sometimes violent events of the past, and metamorphic rocks indicate changes that have occurred, often deep in the Earth's crust. An unconformity may represent an important structural event during the cycle of rock formation and mountain building. After deposition of sediments and the subsequent formation of sedimentary rocks, folding and uplift may occur, to be followed by erosion and the development of an unconformity.

Stratigraphy is the science of putting the Earth's geological history into sequence and also interpreting the events about which the rocks provide evidence. It is based on a number of principles, which are considered in more detail in the chapter on sedimentary rocks. These include the theories of superposition and uniformitarianism, and the dating and correlation by fossils. However, they only provide relative ages for rocks. Absolute or definite ages are determined by using radiometric dating, which looks at the breakdown of radioactive minerals contained in rocks. Only certain rocks, mainly igneous and metamorphic ones, can be used for this work.

Unconformities are breaks in the sequence of rocks, and mark very important events that sometimes cover wide areas. Geologists have on occasion chosen major unconformities to stand as the boundaries between geological time periods. A number of widespread unconformities represent great changes in the Earth's history, and are often associated with mass extinctions.

In many cases, an unconformity is an erosion surface marking a time when already formed rocks were eroded and weathered on the Earth's surface, before a new series of rocks was formed. When the erosion is caused by the sea, a marine erosion surface is created. This is generally relatively level, as with a modern wave-cut platform, when compared with the uneven hills and valleys of an eroded land surface, which may also contain fractures and show indications of chemical weathering. Such distinctive features can be found in different unconformities, suggesting something of their origin.

▶ This low cliff, about 5m (16.5ft) in height, shows deeply weathered Lewisian gneiss lying beneath Torridonian conglomerates, both of Pre-Cambrian age. The unconformity between these two rock types is an eroded ancient landscape with valleys and hills. Here, it is seen as an irregular surface. The Lewisian rocks were highly metamorphosed long before being eroded. These ancient rocks are the oldest in the area and so are referred to as a basement formation. Subsequently, the Torridonian conglomerate, followed by a variety of detrital sediments, were deposited by river systems flowing from the west. This is an angular unconformity, as structures in the Lewisian rocks are at a high angle beneath the Torridonian. The Torridonian conglomerate contains many large fragments derived from the erosion of Lewisian rock.

Angular unconformity, Sutherland.

Very often an unconformity will have features of a marine transgression, when sea level rises and the sea covers more of an old land surface. Basal conglomerates may be deposited on the erosion surface. These possibly represent the pebbles on the ancient shoreline, which is progressively moving inland with the rise in sea level. The pebbles, which eventually form a basal conglomerate above the plane of the unconformity, are eroded from the older rocks lying below. In the case of very widespread unconformities, the younger rocks, formed above the unconformity surface, will rest on rocks of different types and different ages. As a marine transgression occurs, the rocks it first deposits will be older than those formed later, farther inland from the original shoreline.

Frequently the rocks below the unconformity have been folded or tilted. As a marine transgression occurs, and new rocks are deposited on the plane of the unconformity, overstep is said to take place as these new rocks cover tilted strata that are progressively younger in their direction of dip.

▶ A large, rounded exposure of basal conglomerate of the Torridonian, above the unconformity with the Lewisian, is seen here as a 'valley fill' deposit, formed in a water-worn valley in the Lewisian landscape. The conglomerate is composed of fragments of metamorphic gneiss and a variety of associated igneous rocks, set in a sandstone matrix. Beyond the conglomerate is a steep cliff of grey gneiss, rising about 20m (66ft) above the sea. This possibly represents the side of an ancient pre-Torridonian valley in which the conglomerate was deposited. Features such as this emphasise that the unconformity between the Lewisian and Torridonian in this region was an irregular eroded land surface. Today, this Lewisian land surface is still being exposed, and small patches of Torridonian sandstone and conglomerate can be found resting on the Lewisian basement in many places, often some distance from the main Torridonian exposures.

The lower, closer picture shows the conglomerate as a mass of irregular fragments, some angular and some rounded. Larger pieces show typical gneissose banding. The size of these suggests high energy transport (as a powerful agent is needed to move such large rock fragments) and relatively local derivation.

Basal conglomerate, Sutherland.

Sedimentary dykes, Sutherland.

▲ In an unconformity surface, there are often cracks and hollows caused by erosion before the deposition of later sedimentary rocks. Here, a fissure in the Lewisian gneiss has been infilled with dark brown sandy mud of Torridonian age. This contrasts sharply with the banded gneiss. A smaller infilled fissure is seen at the bottom left of the picture. The unconformity surface is here viewed from above and the area shown is about 50cm (20in) across.

► This landscape, which owes its details very much to the underlying rock types, shows two unconformities. The view is from the summit of Cul Mor (849m/2,786ft) looking north. On the summit there is a small exposure of pale grey orthoquartzite of early Cambrian age. This is a marine sedimentary rock formed in a shallow sea. The summit cairn, to the right, is made of this rock. The quartzite rests unconformably on reddish-coloured Torridonian sediments, deposited by river systems flowing from the west and north-west, which are of Pre-Cambrian age.

Two unconformities, north-west Scotland.

The low ground, with many lochs, is formed of Lewisian gneiss, and has the typical hummocky landscape produced by this rock. The gneiss has been dated radiometrically well back into the Pre-Cambrian, at over 2,500 million years old, the isolated mountain rising above this gneiss landscape, Suilven, is composed of Torridonian sandstones with almost horizontal stratification. The surface separating these beds from the Lewisian gneiss is a second unconformity, the bedded Torridonian rocks contrasting with the hummocky gneiss.

Angular unconformity, Berwickshire.

▲ ◀ This angular unconformity shows red-coloured sandstones of Devonian age, dipping at around 10° to 15° towards the sea and resting unconformably on almost vertical greyish siltstones and shales of Silurian age. The time gap at the unconformity is about 20 million years. These views show the striking cliffs that are the southern part of the famous Hutton's Unconformity, first described by James Hutton after he had visited the area in 1788. It is a site of international geological significance, much visited by geologists. Hutton used this unconformity, and other exposures, to work out his ideas about the cyclic nature of geological events. Here, at Siccar Point, Berwickshire, he suggested that the Silurian rocks had been formed, tilted vertically, and then eroded before the deposition of the Devonian sandstones. He realised that there must be a significant time interval between the Silurian and Devonian rocks, because, assuming sedimentary rocks are deposited more or less horizontally, the Silurian strata have been tilted and eroded, but these events do not affect the overlying Devonian rocks.

Angular unconformity, Yorkshire.

▲ Here, at Horton-in-Ribblesdale, a plane of erosion separates, with unconformity, older, highly inclined slates of Silurian age from younger, horizontally bedded Carboniferous limestones. The unconformity surface is a characteristically flat and relatively smooth marine plane of erosion. At this location there is no indication of a basal conglomerate, though in other exposures of this unconformity such a stratum does occur. The limestones above the unconformity progressively rest on younger beds of slate towards the left of the picture. This overstep of the overlying beds, seen when the older series becomes progressively younger in its direction of dip, is a common feature of many unconformities.

Unconformity, Pembrokeshire.

▲ Ancient beaches often form the basal conglomerates immediately above a plane of unconformity. Here, a raised beach has been created in the recent geological past, at a time when sea level was higher than at the present. The raised beach conglomerate and breccia rest on steeply dipping limestone of Carboniferous age, and the eroded surface between the two is an unconformity, with a time gap of over 300 million years. The fragments in the overlying deposit are predominantly derived from the underlying limestone.

► (top) The Palaeozoic rocks on the island of Menorca are frequently folded and tilted. Here, an angular unconformity is exposed on the north-east coast of the island. The erosion surface is relatively flat, suggesting that it is a marine surface. Below the unconformity there are dipping shales and sandstones of Carboniferous age. The limestone above the unconformity was deposited during the Pliocene period, so a time gap of over 250 million years is represented here.

► This unconformity is exposed in the Mendip Hills in southern England. During part of the Jurassic period, the Mendips were above sea level and formed an island around which sediment was deposited. Jurassic oolitic limestone rests on an eroded surface of darker Carboniferous

Angular unconformity, Menorca.

rock. Marine organisms burrowed into the Carboniferous sea bed and these burrows are seen as pale features. The rounded ones are those of the bivalve mollusc *Lithophaga*, and the thin structures were made by polychaete worms. Such a burrowed and bored surface is often referred to as a hardground. The specimen is 8cm (3.1in) across.

Unconformity, Somerset.

GEOLOGICAL TIME SCALE

Era	Period	Epoch	Age (MA)
		Holocene (Recent)	from 0.01
		Pleistocene	1.8–0.01
		Piocene	5.3–1.8
Cenozoic		Miocene	23–5.3
		Oligocene	34–23
		Eocene	56–34
		Palaeocene	65–56
	Cretaceous		142–65
Mesozoic	Jurassic		206–142
	Triassic		248–206
	Permian		290–248
	Carboniferous		354–290
Palaeozoic	Devonian		417–354
	Silurian		443–417
	Ordovician		495–443
	Cambrian		545–495
Pre-Cambrian			4,500–545

GLOSSARY

Amygdale An infilled vesicle (gas-bubble cavity) in igneous rock, usually lava.

Anticline An upfolded arch-shaped structure. The core of the fold contains the older rocks. When the anticline is eroded, relatively older rocks are exposed in the centre and the rocks are progressively younger towards the margins of the structure.

Antiform A fold structure with an upfolded shape. This may be an anticline, which is the correct way up and has the older rocks exposed in its centre, or a synclinal fold that has been upturned.

Asymmetrical fold A fold in which the limbs (sides) are not equal in length. The dip of the limbs on each side of the fold is often different.

Basement rocks Essentially these are rocks of Pre-Cambrian age. Being of extreme age, they tend to be formed of high-grade metamorphic rock, especially gneiss. They form the lowest parts of the stratigraphic column, often unconformably overlain by sedimentary strata.

Batholith A very large, discordant igneous intrusion, usually many kilometres in diameter. Such intrusions are frequently composed of granite, and have a relatively wide metamorphic aureole.

Bed An alternative word for stratum, a layer of sedimentary rock. Bedding planes separate one bed from another.

Bole The red-coloured upper surface of a weathered lava flow.

Boudin When certain, initially rigid, competent rocks are stretched by movement in the Earth's crust, they may become broken between less competent and more plastic rocks, to appear as ovoid masses, often with thin 'necks' between them. These groups of boudins may appear like strings of sausages – the term 'boudin' is derived from a French word for sausage.

Braided river A river channel that is divided into many streams that separate and rejoin. Such river systems occur in both arid and glacial environments, usually where considerable amounts of sediment are carried and deposited.

Buckling This is a form of folding, caused by compression, in which the folded stratum has constant thickness throughout the fold.

Calcite A mineral composed of calcium carbonate, with the chemical formula $CaCO_3$. It is common in minerals veins, and is the main component of limestone and marble.

Clast A fragment, usually in a detrital sedimentary rock.

Clay minerals A group of aluminium silicate minerals, with atoms arranged in sheets. They make up a high percentage of the minerals in some sedimentary rocks.

Cleavage The way a rock or mineral splits along well-defined and predetermined surfaces. Mineral cleavage is related to the internal atomic structure. In rocks, it is best known in low-grade metamorphic rocks such as slate.

Clinometer An instrument used in field geology for measuring the dip of rock surfaces. A clinometer is often combined with a compass for measuring the strike of the beds and the direction of dip.

Competent Beds of rock that resist change are said to be competent. They tend to be relatively coarse-grained and rich in minerals such as quartz.

Conglomerate A coarse-grained sedimentary rock containing rounded clasts.

Country rock The pre-existing rock into which an igneous body is intruded. The country rock nearest to the intrusion will suffer contact metamorphism.

Cross-bedding A feature of many sedimentary rocks in which the layers within a bed are at a relatively steep angle to the base and top of that bed. Cross-bedding is produced in a moving current of water or wind. The overlying and underlying beds may have cross-bedding dipping in different directions, if the direction of the current has changed. Many types of cross-bedding are recognised, and the structure of each type is closely related to the conditions in which it formed.

Crust The Earth's crust is the relatively thin outer layer made mainly of cold rocks. In the ocean regions it is only about 5 to 10km (3 to 6 miles) thick, but it is thicker in the large continental areas, where it is 25 to 100km (16 to 62 miles) thick. The oceanic crust is basaltic in composition, while the continents are mainly granitic.

Cryptocrystalline Having crystals that are so small that even with a microscope they are difficult to distinguish.

Diagenesis The processes that turn soft sediment into rock. These generally occur near the Earth's surface, and although certain chemical processes are involved, no great heating occurs.

Dip The true dip is the greatest angle that can be measured down from the horizontal on an inclined rock surface. Bedding planes dip, as do fault planes and other surfaces. The apparent dip of a sloping rock surface is the dip observed in an exposure such as a cliff face. This may be less than the true dip, depending on the angle at which the exposure is cut through the strata. As well as an inclination, dip has a direction, or compass bearing.

Dip-slip The movement on a fault that is in the same direction as the dip of the fault plane.

Downthrow The side of a fault that is displaced downwards relative to the rocks on the other side.

Dyke A dyke is a minor igneous intrusion. It is a discordant sheet of rock, often dolerite, which cuts across structures such as bedding in the country rock. Dykes may have a very long, narrow, relatively straight outcrop.

Erosion This is the wearing away of rock involving movement. Erosion is the work of rivers, the sea, glaciers, ice sheets and the wind.

Escarpment A generally straight, consistent, steep slope.

Exposure Bare rock on the Earth's surface. This refers to solid rock, not loose rock fragments.

Extrusive rocks Igneous rocks formed on the Earth's surface.

Fault A fracture in the rocks of the Earth's crust that has displacement on either side.

Fold A bend in crustal rocks.

Foliation A surface produced in a rock by structural alteration, often during regional metamorphism. Slaty cleavage and schistosity are two examples of foliation.

Footwall The rocks below a fault plane.

Foreland A resistant mass of rock of great extent, against which large-scale thrust faults move.

Fossil Any evidence of past life that has been preserved in the rocks of the Earth's crust. This includes footprints, trails and burrows, three-dimensional shells and bones, and even frozen mammals. Human artefacts are in the realm of archaeology.

Gneiss A high-grade, coarse-grained, regionally metamorphosed rock, characterised by alternate bands of light and dark minerals.

Graben A down-faulted block between two normal faults.

Graded bedding Sedimentary beds in which coarser particles are at the base and gradually finer sediment has been deposited towards the top.

Granoblastic A texture in metamorphic rocks in which discrete minerals, larger than the matrix of the rock, occur.

Greywacke A sedimentary rock consisting of sand-sized particles held in a clay matrix. Greywackes may exhibit graded bedding, and are thought to be deposited by turbidity currents.

Hanging wall The rocks above a fault plane.

Hinge The part of a fold where the dip changes and bending is greatest.

Igneous Rocks formed by the consolidation of magma or lava.

Incompetent The term used for strata that are relatively more prone to deformation. Shale, clay and volcanic ash are incompetent rocks that are easily cleaved during folding.

Intrusive rocks Rocks produced by the injection of magma into pre-formed rocks below the Earth's surface.

Laminated Very finely stratified.

Lee side The side of a structure, such as a dune or ripple, that is sheltered from the prevailing current of wind, water or ice.

Lewisian Part of the Pre-Cambrian of Scotland. The rocks are generally high-grade gneiss, and are unconformably overlain by Torridonian sedimentary rocks in many places.

Limb The extended parts of a fold structure that lie on either side of the hinge.

Limestone A sedimentary rock containing a very high percentage of the mineral calcite (calcium carbonate). Limestone is often highly fossiliferous.

Lineation A one-dimensional feature in or on a rock.

Lithology Rock type; all the characteristics of a rock.

Mantle The internal part of the Earth below the crust and above the core. The mantle constitutes about 85 per cent of the Earth's volume.

Marine transgression When sea level rises and the land surface is inundated, a marine transgression occurs. Erosion may take place and new sedimentary strata are laid down. A marine transgression usually produces an unconformity between the older rocks and the sediment it deposits.

Metamorphism The alteration of pre-existing rocks by heat and/or pressure, often occuring at great depth in the Earth's crust. No melting takes place during metamorphism.

Metamorphic aureole The zone around a mass of igneous rock altered by heat from that igneous rock.

Monocline A fold having one very short limb; a step-like fold.

Mudstone A very fine-grained rock without noticeable bedding planes.

Mullion Columnar structures associated with the folding of bedding and cleavage planes.

Nuée ardente A very violent gas-cloud volcanic eruption.

Orogeny A period of mountain building, often related to plate tectonics and the moving together of large sections of the crust.

Outcrop The area over which a rock occurs on the Earth's surface. A rock will not necessarily be exposed (visible) over its whole outcrop area.

Overstep The younger rocks overlying inclined older beds at an angular unconformity, the older beds being progressively younger in their direction of dip.

Parallel fold A fold in which any bed involved maintains its thickness when measured at right angles to its top and base.

Pericline An elongated fold that dips away from or towards a central area.

Permeable Allowing fluids, especially water, to run through. A well-jointed rock such as limestone is permeable.

Plate tectonics The theory of how the Earth's crust moves. The crust is divided into six large, internally rigid, mobile plates and many smaller ones. A great deal of geological activity is concentrated where the plates collide and move apart.

Plunge The angle in a fold between a dipping fold hinge and the horizontal.

Pluton A large discordant igneous intrusion, formed at depth in the crust.

Porphyritic An igneous rock texture, having relatively large crystals set in a finer matrix.

Porphyroblast A relatively large crystal that is set in the mass of a metamorphic rock.

Quartz Composed of silicon and oxygen (SiO_2), quartz is one of the most common minerals found in igneous, sedimentary and metamorphic rocks and in mineral veins.

Radiometric dating A method for calculating with considerable accuracy the age of certain rocks, based on the decay of radioactive elements found in minerals in the rocks.

Recumbent When one of its limbs is virtually horizontal, a fold is said to be recumbent. This is an extreme case of overfolding.

Regolith The weathered and eroded rock material formed on the Earth's surface, including soil and river alluvium.

Rift The downthrown area between two faults with parallel outcrop.

Rodding During rock deformation, some minerals, especially quartz, may become elongated into rod-like structures.

Sandstone A sedimentary rock made of grains of medium size.

Sedimentary A large group of rocks formed by the deposition of fragments worn off pre-existing rocks or deposited by chemical or biological processes.

Shale A very fine-grained laminated (bedded) sedimentary rock.

Shield volcano A volcano with gently sloping sides, which is made of mobile basalt lava.

Similar fold A fold that maintains bed thickness when measured parallel to the fold axial plane.

Sill A sheet-shaped igneous intrusion which, in contrast to a dyke, is concordant, following existing rock structures.

Slate A low-grade, regionally metamorphosed rock. Slate is fine-grained and noted for its cleavage.

Sole structure A structure making a cast on the base of a sedimentary bed.

Stoss side The opposite of lee side. The stoss side faces towards the oncoming current of wind, water or ice.

Strata Beds of sedimentary rock.

Strike The direction on a dipping rock surface at right angles to the direction of dip.

Superposition A geological principle that says older rocks lie below younger ones. This is the basis for stratigraphy.

Symmetrical fold Folds with each limb dipping at about the same angle.

Syncline A downfold with younger rocks in the core.

Synform An apparently downfolded structure, which may be a syncline or an inverted anticline.

Tension gash A small crack developed in rock during stretching. Such gashes are often infilled with quartz or calcite.

Throw The vertical displacement on a fault.

Thrust plane A low-angled fault plane along which older rocks are moved, almost horizontally, above younger ones.

Tor An isolated, upstanding, hilltop rock mass, usually owing its weathered nature to structural jointing.

Trace fossil The preserved tracks, trails, burrows, coprolites (fossilised dung) and footprints produced by the activity of an organism.

Tuff Consolidated volcanic ash.

Turbidite A graded bed deposited by a turbidity current.

Turbidity current A slurry of water and sediment that flows from the continental shelf into the ocean. Such currents carry coarse sediment into deep water, and a series of graded beds can be formed by successive turbidity currents.

Unconformity A break in the succession of rocks. This is a 'time gap', where part of the record of geological time is absent. Very different rock types and structures often occur above and below an unconformity.

Uniformitarianism The theory that suggests that 'the present is the key to the past'. Geological processes and events that can be observed today have been happening throughout geological time, and rocks and their structures can be explained by analogy with modern events.

Vein Cracks and faults in rocks, infilled with minerals.

Vesicles Gas-bubble cavities in lava.

Volatiles Substances that readily vaporise and have a low atomic weight.

Wave-cut platform The relatively smooth, low-lying, coastal rock surface, which is washed by the tide.

Way-up Sedimentary rocks, when formed, are the correct 'way up'. Subsequent earth movements may fold them so that they become overturned and the wrong way up. Certain structures, such as ripple marks and desiccation cracks, can be used to determine the way-up of strata.

Weathering The wearing away of rock by in-situ processes such as acidic water, the action of organisms and temperature changes.

Xenolith A mass of rock incorporated in an igneous rock. It is 'foreign', and may be a fragment of country rock caught up in the magma.

FURTHER READING

Here is a short list of books that should prove helpful. Many other references are to be found in each of them.

Collinson, J., Mountney, N. and Thompson, D., 2006. *Sedimentary Structures*, Dunedin Academic Press, Edinburgh.

Duff, P., 1993. *Holmes' Principles of Physical Geology*, Chapman and Hall, London.

Goodenough, K.M. and Krabbendam, M., 2011. *A Geological Excursion Guide to the North-West Highlands of Scotland*, Edinburgh Geological Society, Edinburgh.

Jerram, D., 2011. *Introducing Volcanology*, Dunedin Academic Press, Edinburgh.

Keary, P., 2003. *Penguin Dictionary of Geology*, Penguin Books, London.

Park, R.G., 1983. *Foundations of Structural Geology*, Blackie, Glasgow.

Park, R.G., 2010. *Introducing Geology*, Dunedin Academic Press, Edinburgh.

Pellant, C., 1992. *Rocks and Minerals*, Dorling Kindersley, London.

Pellant, C. and H., 2014. *Rocks and Minerals*, Bloomsbury, London.

Pellant, C. and H., 2016. *Fossils*, Bloomsbury, London.

Roberts, J.L., 1989. *The Macmillan Field Guide to Geological Structures*, Macmillan, London.

Smith, D.G. (ed), 1982. *The Cambridge Encyclopedia of Earth Sciences*, Cambridge University Press, Cambridge.

Stow, D., 2005. *Sedimentary Rocks in the Field*, CRC Press, London.

WEBSITES

The internet can be a source of much useful information, but it can also be rather unreliable. The short list below should be a good starting point, and further searching may prove valuable.

The British Geological Survey	www.bgs.ac.uk
U.S. Geological Survey	www.usgs.gov
Geologists' Association	www.geologistsassociation.org.uk
Open University Geological Society	ougs.org/
Geoscience Australia	www.ga.gov.au

ACKNOWLEDGEMENTS

A wide-ranging book such as this has benefited from the assistance of a number of people. Dr Trevor Morse has made many invaluable comments on some of our photographs, and discussions with him on many topics have been very useful. He has also directed us to various photographic locations. Dr David Morton has looked at a selection of our photographs and his expert comments have proved helpful. Steph Walker helped with a location in northern Scotland, as did Dr Will Watts (Hidden Horizons) on the Yorkshire coast. Roger and Edel Ward have given valuable assistance and provided the photograph from Utah. Emily and Martin Swan (Yosemite and Grand Canyon), Jane and Charlie Hodgman (Gobi Desert) and Pat Tye (Costa Rica) allowed us to use their photographs. John Searby (Picturesk, Whitby) has scanned many of our 35mm slides to an excellent standard. Sid Weatherill (Hildoceras, Whitby) and Mike Marshall (Yorkshire Coast Fossils) have assisted with field locations and trace fossils. Scott Wicking (Scott Wicking Photography) gave us the ammonite from Nepal. Juan Ignacio Ojeda (Publicaciones IGME) supplied us with geological maps and definitive reference books on the geology of Menorca. Grateful thanks to Mr James Beastall and his team at Raigmore Hospital, Inverness, and to Mr Robert Marsh and staff at Scarborough Hospital for their excellent care when one of the authors suffered a serious fracture of her ankle during photography for the book in north-west Scotland. At Bloomsbury, we must thank Alice Ward and Julie Bailey for their expert editorial help and Susan McIntyre for design work. Though we have received assistance and advice from many people, any errors in the book are entirely our own.

INDEX

Page numbers in **bold** type refer to illustrations.